I0026754

The Nurturing of Time Future

Dignity Press

World Dignity University Press

Howard Richards

The Nurturing of Time Future

Dignity Press
World Dignity University Press

Copyright © Howard Richards, 2012.

Published by Dignity Press
16 Northview Court
Lake Oswego, OR 97035, USA
http://www.dignitypress.org
Ref. # 11080

Book design by Uli Spalthoff.
Cover artwork by Brigitte Volz, www.brigittevolz.de
More about this book at www.dignitypress.org/time-future

Printed on paper from environmentally managed forestry. See
www.lightningsource.com/chainofcustody for certifications.

ISBN 978-1-937570-01-9

Also available as eBook (ISBN 978-1-937570-05-7)

Contents

One

The electric melody of her laugh excites star showers in my mind. At each sentence she speaks my heart somersaults; distracted by her blue eyes, I tremble while I cull the fruit; we two are making apricot jam. Sure, I am a rational person sitting at a table in our sunlit kitchen, but she delights me so; her moist fragrance perfumes the darkened orange of my moist handiwork. "Tell me true, my sweetest fascination, why do I love you so much?" And since my love is well-read in classics, moderns, and in the post-moderns, familiar with the Greek, the Latin, and the avant-garde French as well (double doctorate *summa cum laude* from Stanford University) she has the authority to answer me: "Because you are a little bit crazy." I am in la la land where I want to be; hanging out with my dear Plato who famously listed poetry, mystical religion and true love as divine forms of insanity, high on soul-mate and sliced apricots when the gate bell rings; answering the intercom, hearing a young girl begging me, "Can you please give me something?"; answering "Yes;" beginning to rummage to fill a bag for her, an avocado, a bun, banana, asparagus flavor Knorr brand dry soup, another little bun, a plate, a glass, walnuts, apricots, maybe she wants clothing; rehearsing what I will say to her; telling her in my star-showered mind about our neighborhood, about our basic needs system; attempting to explain in words she can grasp that health care, housing, and food are guaranteed here for anybody willing to pitch in to help the neighbors. Only desiring to explain, deciding not to search for clothing she maybe wants, I take to the gate the slim pickings I have already bagged, finding there only our neighbor Carmen returning from work, nobody else; I am too late, she went away; I will have to save my bag and my speech for the next beggar; meanwhile, thank you Luci for giving us the apricots, we are making jam of them; the girl who ran away must have been from another street; the kids in our street know there is no need to beg; if you are unemployed there is always community service supervised by Señora Natalia and the ladies of the Neighbor's Council. Our neighborhood organized itself in practice to refute the economists who teach in theory that without consumers able and willing to buy the products of labor there is no labor, and that is why our economy is stable while theirs is unsustainable. I dreamed that the little girl who rang the bell wanted me to prove to her with my bag that those of us who have more than we need are willing to share with those of them who have less than they need; and please let me dream, don't stop me. You know I know that I do not know, I really do not know, what fantasies dwell in her to me unknown head, but I know also that most of the infinite unknown is merely reality. Dreams with sex appeal made of apricots and sugar make the world go 'round. Let me exaggerate a bit and say it is because of the dreams of lovers, the dreams of martyrs, the dreams of inspired plodders that Chile has evolved to become a moderately social democratic society with a moderately complete welfare state; that it is dreams therefore that make it perfectly possible that at a few times and places, such as for example now and here, it imperfectly actually happens that it is really true, and not just a dream that here on our street and adjoining alley we are able to take care of our neighbors because the government has already paid for the big ticket items like health care for all and massive subsidies for social housing. Thinking things over, calculating that the next beggars probably will be from other streets, I mentally rehearse a modified speech, planning to sound them out on the mutual aid that probably already exists—and probably could be enhanced—on the streets where they live.

But as it turns out the next bell-ringer to punctuate our apricot jamming is El Cholo, whose real name is Osvaldo Duran, who is an alcoholic who lives nearby with six buddies in the humble residence of Pito, whose real name is Santiago Santibañez, who has graciously opened his home to a half-dozen of his comrades-in-drink, who in turn share food, jokes, wine, music, and stories with Pito; and since El Cholo takes the same prescription medication I take, *levotiroxina* 100 milligrams once a day, he comes by to bum some pills off me when he runs out, and I fetch my pill bottle and count out the pills one for him and one for me so that we each get exactly the same number; which perhaps makes a point. Another point drives the lovely melancholy sound of my lovely darling composing an argument, demurring that if dreaming could change the world the hippies would have changed it in the sixties, John Lennon would have changed it in the seventies; Barack Obama now; instead the relentless logic of capital accumulation changes the world, economic calculation, the drive to turn money into more money.

Shattered by her terse precision, bewildered like the dust; I feel she is rejecting my meaning half way through its birth canal while its eyes are closed and its head is in pain; she is blind to the weak glimmering growing light of my vision while it is still hovering like a hummingbird seeking its bearings among honeysuckle vines, hesitantly locating itself somewhere near the middle along a red line running from dreams of love to capitalism. Only let me speak, do not silence me, let me develop the thought that the logic of capital, the logic driving commerce and war and daily life and common sense, is itself a mythic structure derived from the myths that organize modern western civilization; and that a myth is, as Joseph Campbell taught, a waking collective dream. Give me my voice, I hear yours, lend me your ears–let our love be the admiration of the radically other for the radically other, like a bird singing to a rose—let me hover near a body and mind I do not possess wondering what you are thinking stirring the apricots as the sugar dissolves into the water; as you perhaps are wondering what I am thinking as the temperature increases and the clouds of sweet white disappear into clarity. When I try to understand why my stomach sinks when she says I talk nonsense, I quickly see that it is not because her words break the spell of my infatuation; on the contrary, when she fights me my libido eats it up; it is not because I want a platonic soul-mate to complete my own identity; on the contrary, I want a levinasian soul-mate to thrill me with mystery. Nor is it because I fear she is destroying my philosophy, and this for two good reasons, first because I feel only gratitude toward those who correct me, second because my confidence in my main beliefs is so immense that nothing she might say could shake it; it is indeed my opinion that the world's key problem, whose solution would unlock solutions to the others, is that few read books like those I write, which opinion can be regarded either as proof of my madness or as an hypothesis that would be proven false if such books were to become widely read but still misery were to continue unabated the same as now. Now to lay down some initial premises for slowly developing a different viewpoint after this review of false explanations of why I am upset, I first note that overwhelming evidence shows our prehistoric ancestors to have survived, and by surviving made it possible for us their descendants to be born, both by calculating intelligence and emotional intelligence; learning both to forage and to bond; by an ecologically functional social cohesion; so now today I need both to calculate the right ratios of apricots, water and sugar, and to intuit how near her to stand, whether and when to kiss her, where to kiss her, and with what intensity. I affirm second, as a rational articulation of empirical findings and not merely as a theoretical postulate, that society is organized, both on its calculating side and on its sentimental side, by stories; stories shape and give meaning to its rules and institutions; some stories are for good reasons classified as myths, including for example—as sacred stories that illustrate what I have in mind—the mythologies of the native peoples of the Pacific Northwest studied by Bronislaw Malinowski; their myths govern the roles, privileges, and duties of the members of each clan; they govern birth, rites of passage, courtship, marriage, illness, death; rights to fishing waters and hunting grounds, and technology.

I wish I knew what myths organize Señora Natalia, who is next to ring at the gate; who is next to pull me away from the blonde magnet who attracts me; who arranges a funeral whenever an indigent dies in the alley; who mediates quarrels, and calls the police when conflict escalates to violence; who makes sure each young scholar has the books needed for school; who supervises the community service of the unemployed neighbors who work for food. This unemployment policy of our Neighbor's Council is sustainable, unlike the policies of the United States, the United Kingdom, and other less enlightened jurisdictions. Here in our neighborhood we do not borrow astronomical sums from China and then throw them into the economy via tax cuts and subsidies and bailouts to shore up investor confidence, and to shore up consumer confidence, so there will be more investments and more purchases and then—so the story goes—more employment. Our sustainable flows of money to buy the flour neighbors Natalia and Rosita transubstantiate to make bread, using locally gathered wood in their brick ovens, come mainly from the monthly incomes neighbor Gaston and I draw from mutual funds holding shares in multinational corporations. We can interpret the sustainability of food security in our neighborhood as recycling the stories told by Adam Smith in 1776 in *The Wealth of Nations*; thinking of the flour as provided by the upper class, mainly Gaston and myself and our spouses and children, the class that Smith calls the landlords, the class that lives a life of leisure with no need to work because it receives a regular income from land rents, or in our case from corporate profits, and in our case not a class we were born into but one we rose into by teaching in universities and writing books; thinking of the tomatoes and sundry other edibles as provided by the working class, Juan and Braulio, as well as Pito and El Cholo and Hanibal and Mario and Marcos and Carlos and Ramon and Joanna and Teri and Cheno when they are sober; who come into free food from time to time in the course of their work, for instance when they load a truck bound for

Santiago with crates of tomatoes and the truck is full and there are tomatoes left over, and the boss gives them to the workers rather than letting them rot—once we got lucky when a worker gave Natalia twenty chickens to distribute—and another time a dog bit a sheep so badly the sheep had to be sacrificed and—this was El Cholo's idea—the meat was distributed to the houses where there was greatest need; and thinking of Señora Natalia as a social entrepreneur who puts it all together. Natalia earned a diploma in chocolate desserts studying at night; diploma in hand she founded a micro-business with a micro-credit from one of Chile's several clones of the Grameen Bank, but in the end expenses exceeded receipts; now Natalia is on the street, not as a homeless person but as a street-sweeper, doing community service herself while she coordinates it as the delegate of the Neighbor's Council; you see her in the street most days, usually assisted by her sub teen daughter Maria Elena. She has become the neighborhood clearing-house for gossip because the river closes our street at one end and there are no cross streets; so anyone who wants to exit to the outside world—whether by horse-drawn cart, by motor vehicle, by bicycle, or like me and like the majority as a pedestrian—can no more avoid passing our delegate street-sweeper, and to be polite stopping a moment to exchange news, than she or he can avoid being drenched in the perfume of marijuana billowing from the little grey house near the corner. As I walk to the gate to answer the bell and meet Natalia the declining evening sun is modulating the light, bathing our avocado, persimmon and lemon trees, the river bed, and the hills on the other side into prismatic multiples of shades of green and amber; and I am remembering that her husband Juan has developed a specialty in the demolition of old greenhouses, recycling the materials they are made of, and that the two of them are paying off the micro-loan she took out for her failed chocolate dessert micro- business, and that when it is paid off she will probably apply for another micro-credit to take another shot at becoming a micro-entrepreneur. Alone on a bicycle, she politely says hello; hands me three bags, tomatoes, cucumbers, and green beans; politely says goodbye; and pedals off into the dusk. I ask myself, "Why am I having feelings I should not be having?" Why does it upset me so much when my nearest and dearest seems to say that the power of the powerful, the interests of the rich, manufactured consent, and the drive to turn money into more money change the world; but my dreaming does not?" It upsets me so much because it reminds me that I have tried so hard to persuade people to think outside the box with so few results; normal people, left-wing or right-wing, are still normal; their logics touch my buttons because they remind me of my marginal life. When I was seven years old my three year old brother and I used to sit in the kitchen banging saucepans to drown the voices of our parents quarreling in the living room, but they ignored us, shouting at each other just as if we were not even there. My life has continued the same; we are still banging pots; our intended audience still ignores us, it still promotes and believes half-truths leading to dysfunctional consequences, such as maintaining two households instead of one, such as having to celebrate Christmas every year first with my dad's family and then with mom's. The worst half-truth is the myth of natural liberty promoted by Adam Smith in The Wealth of Nations; there Smith was so enthusiastic about the remarkable capacities of money, free markets, and liberty to organize cooperation among millions of people even though they do not know each other that he failed to celebrate the remarkable capacities of kinship, social norms, and religion to organize cooperation among hundreds of people even though they do know each other.

Early modern Europe, where Adam Smith was an eighteenth-century synthesizer of earlier ideas, created something new in history: the modernity that later became global; but neither modern ideas generally nor Smith's ideas in particular are exceptions to a pattern my friends and I see in all cultures at all times: always social order needs culture, culture grows from myths, and myths are driven by dreams.

Nevertheless, our dominant Smithian myths and dreams; tightly welding together liberty, power, and money; fashioning what Max Weber called the iron cage of modernity, a cage traditional peoples can enter but cannot leave; have come to be regarded not just by the majority but also by some scholars who should know better as natural facts and not as imagination-built facts. Immanuel Wallerstein has shown in *Unthinking Social Science* that they are not so much the objects of study as the presuppositions of political science, economics, and mainstream sociology. Immanuel Wallerstein could have added that Smithian ideas and institutions are also the ethnocentric presuppositions of contemporary daily life and common sense; and I myself could have contributed the observation based on my experience that the most ethnocentric debaters win the debates; they are the ones who assume with political science that the world is run by power, and with mainstream economics that people do what they do for money; which winning of debates is to be expected because by definition most people believe what most people believe, and because people agree with speakers who tell them what they already think; but which drives me to despair because it makes me realize how hard it is to make headway against entrenched half-truths like "The world is run by power, the world is run by money, by the money of power, by the power of money;" and which sounds an alarm in my head warning "Howard, you're screwed, trapped again in a fight you can't win!" like when you were pinned to the mat in less than a minute in the first round of the Redlands Junior High School wrestling tournament.

But I did win; I have successfully struggled for optimism, starting with cheering up my pessimistic younger brother. He couldn't stand our mom and moved out; bach'ing it at our grandmother's apartment house in Pasadena. On the wall of his room he scrawled in charcoal, *l'homme est une passion inutile*, a quotation from his favorite philosopher Jean-Paul Sartre at a time when mine was Martin Heidegger, who was the coiner of a number of neologisms gluing small German words together to make bigger German words that when translated use lots of hyphens, like being-in-the-world, being-there, being-question, being-with, being-towards-death, and the goal of all my teenage strivings, authentic-being, a concept I felt more than I understood—feeling authentic-being as magic, as enchantment, as the gateway to the gardens of Heidegger's prose poetry where every flower glows with lights of sublime authority and insight—and feeling authenticity also as a premise from which I could deduce as a corollary being-is-exciting; a corollary I then used to refute my brother when he tried to prove that being-is-boring.

After the fruit boils we stir it and lower the heat to let it simmer in its iron kettle, and since we have only one iron kettle and do not approve of jam prepared in steel, we must wait until it cooks to refill its kettle with new apricots, enjoying a long intermission. She asks if I would mind if she were to play Chopin waltzes on the piano, smiling coyly as if she were unaware that nothing would please me more—except perhaps playing Chopin on my body—but as it turns out she plays only one waltz before retiring for a nap, leaving me trying to imagine how to tame the hordes of books rampaging across our furniture and floors, effectively preventing us from pretending to practice a pretense of good housekeeping. In the earthquake of 2010 the shelves collapsed, heaving books, dishes, clothes and whatever into mixed low jumbles; now, apart from deciding not to rebuild the old shelves—which obviously were not earthquake-proof—we have only gotten around to putting in order the dishes, clothes, and whatever; leaving the books in unwieldy and unintelligible disorder. It occurs to me that I could fish through the chaos to find the books that I myself have written. The first book would be my *Life on a Small Planet: a Philosophy of Value* (New York, Philosophical Library, 1966).

When I wrote my philosophy of value I read "value" as the sovereign word, as the King-word or Queen-word governing all the other words; for "value" means "price" and prices and the system of prices ostensibly govern the economy; while "values" signifies the ideals and the norms that ostensibly govern all conventional behavior, whether economic or non-economic—so much so that educators wisely say that if we do not teach children values, then there is not much point in teaching them anything else. Since I then believed that all knowledge is founded upon experience, and since I then expected a single word "value" to refer to a single thing, namely value; my philosophy had to be a study of experience of value, value-experience. First I had to answer my brother's objection that there is no value, which I did by arguing that though you may feel blue now sooner or later a beautiful

21

value-experience will happen to you and then you will change your mind and decide life is worth living after all. Anyway, it happens to me; therefore it exists. Then most of the book answers the question when value-experiences occur, which might sound like a question to be answered by empirical research, but I was then hypnotized by Martin Heidegger's procedures for discerning philosophical knowledge that can be established prior to and independently of all science. Since Heidegger did philosophy quoting the German romantic poet Friedrich Holderlin, I copied his method quoting the romantic American poet Kenneth Patchen. Since Saint Thomas Aquinas refers to Aristotle as "The Philosopher," I copy Aquinas calling Patchen "The Poet." What follows in practice is that we should go barefoot more often. Authentic value-experience, authentic art, happens close to nature, in being real sunflowers not fake locomotives.

The *Evaluation of Cultural Action* (London: Macmillan, 1985) stars peasants eking out a living on poor soil in the rainy south of Chile, at a time when General Augusto Pinochet was the grim chieftain of an all-pervasive violent capitalist revolution. A poignant story about family and friends bonding and benefiting under rain and repression frames a sober dialogue about methodology, starring Paulo Freire. Freire's life was a failure in the sense that he devoted it to persuading people to think outside the box, but at the end of it most people still thought inside the box; nevertheless (I like to think my life is like his) instead of succumbing to pessimism Freire enthusiastically practiced a methodology for transforming the basic cultural structures of the modern world, structures one might name—heroically simplifying Max Weber—as calculation, capitalism, and bureaucracy; and with these words, dear reader, I let you in on a secret: the secret is that "the box" and "the basic cultural structures of the modern world" are synonyms. "Basic" means governing meeting basic needs, like the need for food; "cultural" means developed in upbringing from play, music, stories, learning to cooperate... culminating in ethics (norms, rules) governing action; "structures" are institutions made of rules (the roots of rules are cultural and biological); "modern world" is roughly what Max Weber described.

My two volume (personal) *Letters from Quebec: A Philosophy for Peace and Justice* (San Francisco and London: International Scholars Press, 1995) maps out a game plan for humanity and the biosphere. Its historical examples (and game plan) of cultural action,—that is to say of changing basic structures by changing culture—acknowledge certain limitations; they acknowledge that even though culture is socially constructed, deconstructible, and reconstructible; even though culture is made possible by the spirit of play (thank you Victor Turner) and by the spirit of storytelling (thank you Marcel Mauss); nevertheless even the most creative cultures survive only if they can pass reality tests imposed by ecology, by human nature, and by warfare; only if the cultures avoid destroying their habitats; only if they successfully organize the metabolism of society, its exchange of matter and energy with the environment (thank you Karl Marx); only if they do not rely on nonexistent motivations but instead successfully tap the energies of biological behavioral tendencies (*Triebe*) that actually exist (thank you Melvin Konner); only if no military disaster ends the culture by massacre and/or by imposing an alien tyranny that prevents the young from learning the ways of the ancestors.

While the examples in *Letters from Quebec* are Plato, Thomas, Kant, and other great western philosophers, interpreted as having been doing cultural action whatever they might have thought they were doing, *Understanding the Global Economy* (Delhi: Maadhyam Books, 2000; revised edition Santa Barbara; Peace Education Books, 2004) turns to economists. Since we humans make ourselves predictable with culture —myths, promises, rules, institutions...; whenever economists explain our conduct, saying X causes Y, they inevitably refer to myths, promises, rules, institutions... In particular economic explanations require the economy's legal frame, whose most important historical source was the Roman *jus gentium*. Thus Max Weber says economy requires law to make the consequences of decisions *kalkulierbar* (calculable).

My review of my books now pauses on hold, to let me broadcast a short confused report from my emotions. Pot-banger at first, with my young brother as the first of my collaborators, I have never ceased feeling I had something very important to say to people badly needing to hear but not at all interested in listening; today I sincerely feel that if only more people understood cultural action to change basic structures peace and justice would be feasible. At first "frightened child," then an academic watching street beggars, the news from the Middle East, global warming, thinking all this is "mistake," caused by ignorance not by nature, greed, or sin; I feel I must work to correct the mistake, and at the same time fear I might really be what some say I am, unintelligible, legitimately ignored, because by definition when words are unintelligible the question whether they are true or false does not arise. I am hopelessly in love and deliriously happy. I often wonder, in my pleasures, whether apricot contentment slows my work. I fear I should work for world enlightenment 24/7. And then Uncle Sigmund Freud's theory of psychic health comforts me. I believe Freud meant—certainly his disciple Erich Fromm meant— that work, play, and love enhance each other, so that without love and play work withers.

When *aymaras* of Bolivia hold a *minga* to start the agricultural year, coming together dancing and singing to clean out their irrigation canal, and then conclude the work/fun with feasting and drinking, I think they too motivate business with pleasure.

Back to work: *Dilemmas of Social Democracies* (with Joanna Swanger, Rowman and Littlefield, 2006) examines struggles to democratize economies in six countries and at the World Bank. Studying Spain, Sweden, Austria, South Africa, Indonesia, Venezuela, and the World Bank last, the book shows how neoliberalism triumphed when social democracy failed, and—note this—that its failure was inevitable given the basic cultural structures. When you think and act inside the box, higher wages and higher taxes mean lower profits and less investment. We show that this principle (wages and taxes up, investment down) follows from the legal and ethical rules of the system, even though in the empirical data the operation of this principle is sometimes temporarily masked by cross-cutting factors; and we show how because of it—because high wages and taxes drive away investment—most efforts to democratize economies attempted so far have not been feasible. Our findings do not make us pessimists; they make us cultural activists. To say it is not feasible to build green and fraternal futures within the confines of the basic rules and the common sense of our so-called civilization, is not to say that green and fraternal futures cannot be built at all; it is to say that to build them it is necessary to work at deeper levels; at the levels of worldview and religion, at the levels of infancy and early childhood, at the levels where biology intersects culture. "Cultural action" used instead of "political action" or "social action" names those deeper levels. My next book was a study of Rosario, Argentina. It reports on that city's practical steps to gradually transform the basic cultural structures.

The right words for making transformation intelligible grow from right practice. They are not meaningful without use. They grow from language-games one cannot fully understand without playing them, like for example the *Sprachspiel* Señora Natalia plays so frequently using the word "cooperation:" Gaston and I cooperate buying flour, Natalia cooperates cleaning the street, Juan cooperates gleaning tomatoes, Lucia cooperates as teacher's helper in the local elementary school; for the five of us "cooperating" is a ticket of admission into fraternity, into the extended family of the neighborhood. I want to stretch language "up" and "down;" up to articulate the framework of a Game Plan for transforming the global economy; down to communicate in the micro-contexts where the rubber hits the road, where every neighborhood and family and city and sport and church and school and workplace and hospital and hangout for drug addicts and jail and political movement has its own *argot.* I can show samples of local *argot* from Rosario. You have to be there and walk the walk to understand the lingo of "social militant" or "urban farmer" or "sexual diversity officer" or "Municipal Bank Foundation" or "watchdog of the participatory budget."

Having been for so many years so ignored while being so irrationally convinced that the fate of the world depends on my ideas, I have had time to ponder how to communicate my message better.

I have decided to write a short book about everything, addressed to educated people of good will who see the need for paradigmatic change. Its method will be to say what I feel needs to be said and to do my best to explain the reasons and motives that make me feel I need to say it. I will propose a philosophy in the old-fashioned sense of articulating a synthesis of the social and natural sciences that provides an overall framework for deciding what to do.

But now my lovely darling has finished her nap and resumed playing the piano; I shall postpone trying to organize our books to another time; now instead of Chopin she plays Haydn's Sonata in D Major; now it is time to move the fruit and the kettle from the kitchen to the living room while I listen to the magic of her fingers. My own fingers start another batch of apricot jam; and soon it will be time for them to wash my feet. Once she politely suggested that the odor of my lower extremities was suboptimal, and since she gave me that hint I have been washing my feet with deodorant soap and spraying them with cologne before going to bed. And then it will be time for more star showers in my mind, waltzes in her fingers, somersaults in my heart; and I shall fall asleep beside her and dream of orange fruit and sugar, of sharing pills with El Cholo, of a bird singing to a rose; of neighbors on bicycles bringing tomatoes, cucumbers, and green beans; of iron cages, of philosophers and economists, of *aymara* women wearing black hats cleaning irrigation canals and drinking beer.

Two

I dream of making points clear to Messrs. Gates, Soros, Gaga, and Winehouse. I am not rich or famous, but I am not nonexistent either; I exist here in my small corner of the world; I have my own point of view. I do not believe the central problem is that you and the other rich and famous people have power and influence while I and other poor and unknown people do not. I do not think I need to persuade you to support worthy causes. You are not against ecology or peace or social justice; so it is not a matter of persuading you to change sides and start working for the good of the biosphere and the world's people; it is a matter of rethinking thinking, changing paradigms, finding social change strategies that work. I only plead for a chance to explain to you how to turn the tide; to propose a Game Plan with two sides: on the one side methods for getting out of the lockstep of compulsory obedience to the systemic imperatives of capital accumulation, so well described by Ellen Wood and David Harvey; and on the other side methods for detecting and enhancing in any given milieu the cultural resources (or social capital) available to organize cooperation and sharing there—two sides of the same Game Plan because enhancing not-for-profit public and private institutions builds a plural economy loosening the death-grip of the systemic imperatives. Here Erich Fromm's ideas imply more good news: while our buying-and-selling way of life tends to reduce people to personalities-for-sale while reducing the rainforest to opportunities-for-profit; when we join together in communities of resistance to overcome this fragmented and fragmenting way of life (uniting for some practical purpose like cooperative day care or community supported agriculture), we feel joy and make friends in the short run at the same time as—by decreasing addiction to profit—we contribute to changing the system in the long run. In my own small way I try to be a catalyst organizing cooperation for non-commercial motives in my family, in my neighborhood, in the schools where I work, in my church, in my political party—uniting people for practical purposes. I do not claim to be more obscure than the millions of other obscurities who are resisting fragmentation by building social capital, but I do claim to have a useful theory justifying what we do. If I were famous enough to get in their doors I would suggest to Messrs. Gates and Soros, Ladies Ga Ga and Winehouse, that ethical norms are causes; changing the norms changes history. Martin Luther King Jr. and Mahatma Gandhi gave us examples. Martin Luther King Jr. drew his strength from the black church tradition (as James Cone has shown), a tradition with roots in African conviviality, in the spiritual consolation of slaves, in keeping joy and hope alive in times of low wages and unemployment; while Mahatma Gandhi was able to think and act outside the box (as Joanna Swanger and I have shown in our book on him) because beginning in his childhood he lived and breathed an indigenous *dharma* that his reading of anti-modern Europeans—like John Ruskin and Leo Tolstoy—only complemented; much as the philosophies of the Boston personalists complemented the gut intuitions of King. One coming from Africa and Christianity, one coming from India and Hinduism; both show how filling in the ethical gaps in the modern western economic way of life is often achieved by recovering ancient wisdom, not-western or not-modern.

My only desire is to explain wisdom—not to explain why I choose to be in this world but not of it, keeping my heart in la la land with Plato where truth and beauty have an eternal affinity for each other (since explaining myself would chiefly interest the few who wonder why I am a misfit); but to explain wisdom as a path from here where we all are in an unsustainable dysfunctional world to there where we all want to be in a world that works; and to start I will explain wisdom as a remedy for two things that at the present time are not working: (1) our efforts to prevent crime, and (2) our efforts to rehabilitate criminals; beginning by hearing our neighbor Carlo who says his house is marked to be burglarized by "the same old crazies

31

and druggies." Somebody has recently spray-painted two large capital letters on Carlo's front wall, a large backwards E and a large frontward B superimposed on a cross-hatch pattern that Carlo believes is a map showing the thieves how to find their way around the interior of the house after they break in.

Since the words "ancient wisdom"—which according to me name the solutions to Carlo's problem—form a common phrase, almost a cliché, while "modern wisdom" is an odd combination, which seems false, like a cat pretending to be a rabbit; one can perhaps discern a truth in the poetic hyperbole of the implausible assertion of Jacques Maritain that modern Europe is kept alive by the prayers of the faithful in the monasteries. In *Habits of the Heart* Robert Bellah and his co-authors make a similar point more soberly: the USA does not—and it could not—function using only its modern and dominant language, business-talk; it also employs the ancient languages of religion and civic virtue, as well as several versions of therapy-talk. From Maritain and Bellah I derive my first point: ancient wisdom is not gone; it is still here with us as a substrate underlying modernity, making modernity possible by tempering it; and my second point is that to the extent that we have some limited success in rehabilitating criminals it is largely because of their religious conversion, or—as they say in the prison ministry in Chile—their *caminando el camino*. My third point is that my second point shows that my point of view, Martin Luther King's point of view, Mahatma Gandhi's point of view, are not absurd, not legitimately to be ignored—because the soul of the criminal is in practice the indispensable agent of his or her rehabilitation, because the partial solutions we already have to Carlo's problem in fact rely on wisdom. Rehabilitating the authors of the *graffiti* on Carlo´s wall would have been duck soup for the wisdom of Saint Francis, who converted a wild animal—a wolf who terrorized the village of Gubbio devouring alike dogs, women, and men—by preaching to it the love of authority and the authority of love, and then—so it says in the *Little Flowers of Saint Francis*—cutting a deal with the wolf on the following terms: "I promise thee that thou shalt be fed every day by the inhabitants of this land so long as thou shalt live among them; thou shalt no longer suffer hunger, as it is hunger which has made thee do so much evil." From this miracle performed by Saint Francis of Assisi I derive three precepts, three rules for cultural action, three methods for enhancing the available cultural resources in any given milieu: (1) Make a strong emotional appeal, (2) Affirm authority, (3) Solve the material problem; and now in the light of these precepts I will consider the attitudes of delinquent youth in Chile today, the kind of youth who say to their high school guidance counselors, as one whom I interviewed did say to his, "I can make more money than you make by stealing and pushing drugs," the kind of youth who grow up expecting to spend part of their lives in jail, because their parents and most of the people in their milieu spend time in jail. Apart

33

from some fascinating individual and subgroup differences, I found among the kids I interviewed that although they are defined by the school system as at-risk, as vulnerable, as underachievers, they define themselves as happy and normal young people, who love their families and friends—the same families that teach their children how to steal, the same friends that hang together spraying *graffiti* on walls—who enjoy even school because it is a place to be with their friends; who thrive on all-night parties with lots of dancing to sensual reggaeton music, lots of beer, lots of sex; and who in normal hours in between the heavy pleasures of weekend nights spend most of their time watching TV, playing soccer, flirting, hanging, and chatting on Facebook.

My minor research on the attitudes of minor authors of minor crimes shows their approach to life to be similar to that of the psychotics Uncle Sigmund described as deeply attached to their psychoses, because, in one case as in the other, a strong emotional appeal drawing them into conformity with socially defined normality would have to be extremely strong because strong basic emotions already attach them firmly to the way of life they already have. I know it can be done because I know pastors in Chile and elsewhere who keep reformed delinquents so busy with healthy fun that they do not have time to get drunk and pregnant even if they want to. But now instead of entering into the general topic of how some cultures succeed and some cultures fail in the sacred task of

convoking angelic choirs of higher spirits whose singing deafens young pre-delinquents to the siren songs
of orgies and easy money, let us turn to the topic of
affirmation of authority. Starting my investigation of
this topic here where we are now when it is, my data
show that authority is affirmed by the pastors who here
and now save souls from perdition by repeating over
and over again a single central message: Not my will
but thine be done, on earth as it is in heaven. These
words ring strangely in the modern world, for there
is no way to reconcile them with personal autonomy
as the supreme principle of morality; they express
a categorical rejection of modernity's basic cultural
structures, even though today in the cities, suburbs and
countrysides throughout the planet they continue to
be spoken, sung, and prayed by millions, by billions;
let thy will not mine be done still endures as song
and substrate contradicting and accompanying the
individualistic mythology now organizing the global
economy; these ancient words coexist with modernity; these ancient words still express the meaning of
wisdom, or rather they would express the meaning of
wisdom if the powers that dominate the manufacturing
of knowledge and writing of curricula would allow the
people to remember what wisdom used to mean, what
the people still instinctively feel it to mean, and what
it means today for scholars who remember Plato, who
was the great spring at the headwaters of the rivers of
western classical tradition, because in Plato the definition of wisdom is the harmonizing (not suppressing)

government of the entire soul by that part of the soul which has the word (the *logistiche psuche*), because in Plato the *logistiche psuche* governs pride and appetite; wisdom is a form of obedience to authority, it is obedience to the word (*logos*, commonly identified with the divine *Logos*). Speaking from a naturalistic and scientific point of view we can say that producing an ecologically sustainable social cohesion has been the central challenge not only for Plato, but for sages of all times and places, some of whom like Plato have identified the Higher Power that brings order to souls and communities with the Divine Word, while others have named the Higher Power some other way—as ancestors, as spirits dwelling in the forests, as pantheon, as patriotic valor, as love, as ultimate concern (Tillich). But there are times and places where a generalized decline of legitimate authority sets in, often because authority loses its legitimacy when it fails to deliver security and material prosperity (Habermas); where anomie (normlessness, social disintegration) destroys humane government of souls and communities starting a downward spiral into crime, mental illness, drug abuse, senseless violence, children growing up with no moral compass to tell right from wrong, depression, and disorder. Since the remaining alternative—neither the affirmation of obedience to cultural norms nor decline into disorder—is the establishment of a semblance of order by force, cultural action to enhance existing cultural resources (to make them less dysfunctional, more functional) saves us from not one but two terrible alternatives: from chaos and from right-wing reaction; cultural action is thus an intelligent progressive response to the conservative abhorrence of chaos forever demanding the establishment of peace by violent means: first by "law and order," then by repression, and finally by outright dictatorship. As Saint Francis knew that to tame a wolf you have to feed it, so a third of three general rules for cultural action today is: solve the material problem. (First: a strong emotional appeal; second: affirm authority.) A material aim of criminal rehabilitation is to re-insert people in the labor market with good jobs at good pay, but this aim is routinely frustrated because the number of jobs is smaller than the number of people who want jobs. I have been explaining in installments—partly in the previous chapter and partly in other writings and PowerPoint's—why recovering ethical wisdom (thinking outside the box) solves the basic problem of too-few-jobs; and here starts another installment. I witnessed a wonderful training workshop in a poor part of Africa for 257 unemployed men and women; some were young some old, some moderate to severe alcoholics, some teetotalers because they were Rastafarians. The poor learn to organize themselves, following the steps of the method of Clodomir Morais; the participants put each theoretical lesson into practice, organizing themselves into work teams and then submitting bids on community service projects that have been lined up and "scoped" in advance by the teaching crew. The newly created work teams analyze the jobs; they prepare work plans and budgets; then they submit bids; and then they sign contracts with the teaching crew, which is also a technical crew qualified to evaluate the work done, for—in the case of the organization workshop I witnessed—building a day care center for children, remodeling an old peoples' home, planting trees, cleaning up the playground of a school, planting community gardens...; and then when a job is done the team submits an invoice and is paid (at market wage rates) as per its contract.

Notice first that the unemployed found employment without depending on effective market demand, by doing community service, without out-competing other job seekers, without throwing others into unemployment, doing work that was not commercial; then imagine achieving full employment by relying not on a single logic—the logic of sales and profits—nor on two, but by relying on several diverse logics. Notice second that the participants learn to work together in ways that could subsequently be utilized in many ways (in the world as it is and in the world as it might become) in commercial and in social enterprises, as has indeed happened, for graduates of organization workshops have gone on to organize in the private sector, in the public sector, and in "third sectors." Notice third that the wages of the participants—paid by a foundation with both private and public funding—are a channel for sharing the wealth, for recycling funds into the lower strata of society. Changing the basic rules of the game of human life, rules about buying and selling, doing

38

things with money, working and being worked for; as is being done by organization workshops and by thousands of other innovations in economic solidarity found all around the globe; has seemed to me to be, since I was eleven years old, a matter of playing clay, as I did when I would leave my divorced mother and my four-years-younger brother in our shack in the slums of Barstow, California, to wend my way to an elegant residence surrounded by gardens, an oasis in the desert, to meet my school chum John Hinshaw and his sister Susan to play clay, a game consisting of making whole cities out of modeling clay; making tiny women and men, their children, their buildings, their vehicles, and of course their gardens, for it was against the rules to make people without gardens (because if there were no food what would the people eat?); and when the top of the Hinshaws' patio table was entirely covered by a clay city, we would cross the lawn to make a "colony" on a large flat rock amid rambling roses and twining ivy.

Back at the shack in the slums I played a game with my little brother, making cities from stiff paper oblongs punched with small square holes. We had an enormous supply because our mother taught at a school that owned a primitive computer. She was able to bring home for us the stacks of used computer punch cards we used to make first a house of cards, and then another and another, and then a school and a fire station and a grocery store, until—so carried away by an orgy of world-making that we forgot to eat and drink, wanting to be God but not daring to admit it even to ourselves—we covered the entire floor of the room with a used-computer-card city; which strongly suggests some things about our indulgent mother, since because we only had two rooms, and since our mother never complained when the floor of one of them was covered with play buildings made of punch cards, she probably had read Jean Piaget, who made the discovery that children are biologically programmed to make up games with rules and to play them; and she certainly had taken to heart the piagetian slogan "play is the work of the child."

Later our games imitated capitalism; elaborating on the board game Monopoly we supplemented the simple version for sale in stores with complexities we invented ourselves like corporations with shares that could be traded and which paid dividends when you passed "Go." Piaget also found that children are programmed to form groups, a finding I confirmed by being a full-fledged member of two child-organized gangs; first one called "Four Bafflers" (although it had ten members) which arose out of the dense mists of nothingness and took on mythical form and liturgical substance in the fertile imagination of latency in the town of Fontana, California, where we lived before our mother left our father and took us to Barstow. We inducted our new members trespassing into an old warehouse filled with bales of textured cardboard (waiting to be made into egg cartons) that we had rearranged to create a labyrinth of tunnels, through which we led blindfolded novices into an inner chamber also made of bales of cardboard, where they were solemnly initiated into "Four Bafflers" by candle light. I am consequently a person with a not inconsiderable background in game-invention. I think this fact about me will help you to understand why I sink into such despair when I watch the evening news on television.

I watch Barack Obama, whom I had earlier watched saying he would raise spending and lower taxes for the purpose of increasing demand and therefore sales and therefore profits and therefore employment; now saying he will lower spending and raise certain taxes for the purpose of lowering the rate of growth of government debt; followed by Representative Boehner, the Speaker of the House, who says the president's current plans for taxing the rich will increase unemployment; followed by Robert Reich, Secretary of Labor under Bill Clinton. When Reich says Obama is fighting fire with gasoline, taking money out of the economy when he should be putting more money in to stimulate growth and employment, it makes me want to curse the box and scream: "Why can't they invent new games and play by different rules?" Then my mind flashes back to Barstow when I was in Fifth Grade and a member of the Kreeper Gang (named after Ray Kreeper its leader) when I first encountered a complexity I have often encountered since: that of being both anti-authority and pro-authority. The point of the gang was to break the rules of the school. But our gang, like the football hooligans studied by my Oxford tutor Rom Harre and

41

co-authors in *Rules of Disorder*, had its own rules, and it was essential to respect the mysterious authority of the gang, as among the pygmies studied by Colin Turnbull it was essential to respect the spirits of the forest. Homer—I wish I could remember his last name—told me he was once caught pissing within a hundred feet of our hideout—which flouted our most important rule—and he had been whipped for it. He took off his shirt to show me the whipping scars on his back, proud to bear on his body the proof of the Power of the Rule. That meant The Power of the Gang, which for me meant Protection. I got all the way through twelfth grade and graduated without being beaten up mainly by having tough friends like Ray Kreeper.

Defying authority was my specialty because my role in the Kreeper Gang was to dream up ways to have fun breaking rules, like wet-paper-towel-skiing, thus justifying a membership in the gang that could not be justified by any ability to defend my comrades with my fists. To do wet-paper-towel-skiing, we would go into the Boy's Bathroom, take down wads of paper towels from their dispensers, soak them in water, and then throw them on the floor. We would get a running start, jump on the wet towels and slide across the floor until we crashed into the wall --an activity we would continue until we were caught and punished; the punishment phase being a further fount of fun and also a producer of prestige for the glory of the gang; as was marble-dropping—an activity which began when we were all seated at our desks in straight rows, and I would unobtrusively drop a marble on the floor; signaling the gang members unobtrusively to drop marbles on the floor too, leaving the teacher confused by the noise and unable to discern who had dropped one and who had not. Having been such a goof-off at age eleven, I might now trumpet "Question Authority" everywhere, but instead I have grown up to echo John Ruskin's call for respecting legitimate authority and for opposing it when it ought to be opposed "... loyally and deliberately, not with malicious, concealed, or disorderly violence."

The need to be both pro-authority and anti-authority, both respectful and critical, both for and against rules, follows from the need to change from today's set of basic rules to tomorrow's set of basic rules. The system-changing process is a rule-changing process because systems are made of rules, because institutions are made of rules; and since rules require authority (*Rechtfertigung* in Wittgenstein's German), if contesting the old rules dissolves authority, then people will not obey the new rules; from which it follows that if you (mistakenly) view the system-changing process as a struggle for power, overlooking Karl Marx's main point that the basic rules of the capitalist game—mocked by Marx as Freedom, (formal) Equality, Property, and Bentham—ought to be replaced by rules serving everybody not just a few (to the extent that it might be said that a thinker so comprehensive and profound as Marx had a single main point); and consequently—having overlooked what I am claiming to be Marx's main point and not thinking of rule-changing but of power-destroying—you adopt as a strategy for social change the undermining of an oppressive culture by undermining the general habit of obedience to rules, then the result inevitably will be—if you succeed—what we saw so often in the twentieth century: socialisms that do not work.

Standing here and now on our blue mother earth spinning into a new century but still littered with the wreckage of many failed twentieth century socialisms and social democracies, and having mentioned one distinguishable type of left-debris, the anti-authoritarian type; I go on to add overconfidence in theory as another way to fail. When I think anti-authoritarian I see Rolling Stones: sure they fled Laborite England to avoid taxes; sure they recorded in Alabama to avoid the musicians' union; but they identified as left because they defied rules. When I think overconfidence I have in mind neglecting local cultures and relying on a supposedly universal science of revolution. Nevertheless, I still insist on going back to old Marx to understand the global forces that shatter local cultures.

Yes, that is why Adam Smith and Karl Marx are both unavoidable, why they frame both our dreams and our nightmares. These four—Freedom, (formal) Equality, Property, and Bentham—constitute what Marx mocks as a bogus Eden of natural rights; these four govern the Master Game, the game of buying and selling; they mirror precisely Adam Smith's "natural liberty;" these four nail the box; they organize our global economy; these four create both the pretexts and the underlying dynamics of our wars. Each is free to buy or not to buy, to sell or not to sell; all face each other as formal equals in the market place; each disposes only of her or his own property, which is their labor power in the cases of workers with nothing else to offer for sale; each looks only to her or his self-interest (self-interest is a principle of Jeremy Bentham, who analyzes all human action as seeking pleasure and avoiding pain). Follow these rules of natural liberty, says Adam Smith. Yes, that is the question, the unavoidable question, to Smith or not to Smith. Some of us misfits answer it weird, dreaming, making new myths. I want to tell Gates, Soros, or Lady Gaga or Amy Winehouse. But I fear that if they came down from Olympus to glance at me they would say, no, forget him, he does not exist, he does not have a point of view. Because when you live in the box, which is where I believe they live, you see only a limited number of possible points of view, and mine is not among the possibilities, nor is Mahatma Gandhi's, nor Martin Luther King's, nor John Ruskin's, nor Hjalmar Branting's. I want to preserve and improve the "cultural resources" of any given milieu, because like Hannah Arendt I do not want a world where governing norms are gone and "anything is possible." One can understand why Arendt confuses avoiding "anything is possible" with not changing the Roman law tradition our basic rules come from, because Europe was saved from chaos by that tradition more than once, and it fell into chaos when Hitler and Stalin ditched it. But those basic rules of our life-game today make everyone's bread and butter depend on the confidence of investors. So the Game Plan to change the world is not to boil investors in olive oil or in vats of vanilla extract; it is to change the rules to free humanity from what the Grenoble School calls "regimes of accumulation" where all the elements of culture—from designer jeans to military hardware—are pressed into the service of the logic of capital accumulation; it is to encourage sustainable living. I realize you still have an unanswered question: "Who is Hjalmar Branting?" Here is the answer: Hjalmar Branting became in 1920 Sweden's first socialist prime minister, but since the socialists had only a one-vote majority in parliament, they decided not to implement their program until they could get a broader social consensus behind it, emphasizing meanwhile training workers in the management of cooperatives and unions to prepare them to assume the government of the nation at a later date. Son of a professor, himself an astronomer and a gymnast, Branting believed in *Uppfostran. Uppfostran* in Swedish means roughly "self-improvement" in English. Like Plato Branting promoted personal self-improvement as part and parcel of improving society.

Remembering Branting and Swedish social democracy brings tears to my eyes. For a time it was the world's model. But in my brain I know the Swedish Model fizzled; and that is why we need to go back to the drawing boards; and that is why I cannot say, "Here

I am, another center-left progressive!;" and instead I must insist on taking the analysis to deeper levels—to the level of the myths that organize cultures, to the level of primal dreams—and that, in turn, is why I fear that in the eyes of Messrs. Gates, Soros, Gaga, and Winehouse my point of view does not exist; it is too new to be on the charts. Seeing the issues as deeply cultural leads to using participatory methods; it leads to listening to detect what is meaningful to people and to discern their energies; as a community organizer does when setting up a Neighborhood Crime Watch —calling a meeting, hearing what people say, writing down their words in glowing Technicolor on butcher paper taped to the wall. I wish I could explain these ideas more clearly. I call detecting meanings myth level work and discerning energies dream level work; and I want to say we need both to overcome the forces that stopped social democracy and imposed neoliberalism on the world.

At the myth level, I have decided to write a short book about everything, addressed to educated people of good will who see the need for paradigmatic change; at the dream level I dream of Carlo.

I dream of Carlo's musical talent making him a Generative Person, a source and example of culture-shift on our street and in the adjoining alley. Generating a culture-shift requires energy, and in Carlo's case the change-energy is clear: he craves an audience; he lives for applause. I dream of Carlo performing for our Having Fun Being Good dances, sing-alongs,

Christmas parties and sundry amusements—the good times that have earned us a reputation as the Home of the Singing Drunks.

We are proud of our reputation because we know that while descending sensually into a bottomless pit of dysfunctional behavior is easy, and that self-discipline is praiseworthy; we know too that the masses need Gaga; and we know too that persisting faithfully in functional behavior without pleasures is impossible. Millions of years of primate evolution have programmed seeking Gaga into human brains and bodies. Consequently a population capable of being good without having fun is so unlikely that any Game Plan worth diddly poop must activate old ways and create new ways to unite ethics with positive emotions. Wherever you are, you need to tune in to the energies that move people in the neighborhood, for you do not know their forms *a priori*, but first of all—even before you tap Gaga and Go Go to fuel transformation—you need to listen to people, to notice how their words mesh with their feelings and their actions, to write down exactly what they say, so that when you communicate with them they will understand you because you are speaking to them in their own words.

Three

I feel that the young people of Cairo are counting on me to save them. Now they have overthrown their dictator; it is March of 2011; they want decent jobs at decent pay, peace, freedom, democracy. Unfortunately their dawn is dawning at twilight, when decent jobs are scarce all over the world, when social safety nets are unraveling even in the old democracies. Today they want a yesterday that is now vanishing in Europe. My fear that the same old untenable but dominant ideas that are unraveling the rest of the world will unravel Egypt is fed by seeing lines of unemployed men leaving Egypt to seek work elsewhere on Cable News Network and British Broadcasting Corporation television. Then come interviews with experts saying Egypt needs development to create jobs; calling for improving Egypt's global competitiveness; declaring that political and economic stability require a secular state following international models, neutralizing the local radicals, hermetically sealing off the corridors of power to exclude the Muslim Brotherhood and other groups regarded as Islamist, tribal, socialist, radical and/or extremist, doubting that the Egyptian military will be able to perform the delicate balancing act needed to reassure the financial markets that Egypt's new democracy will not become populism. The experts on television say politely but clearly, in discourses studded with euphemisms decorating but not disguising their meanings, that if appearances of people-power cannot reliably be alloyed with realities of money-power, then the military can be counted upon to impose on Egypt now as we speak dancing in the streets to celebrate the end of Hosni Mubarak's thirty-year-long de facto government, yet another de facto government—which would be one, but only one, of the outcomes I feel the young people of Cairo are counting on me to save them from. I have been a faithful student of the philosophy of Ludwig Wittgenstein; I have made with him the transition from his early *Tractatus* to his late *Untersuchungen*; from the logic of universal science to the anthropology of diverse language-games. Since I do not believe there is a Single Truth in Science in Marx or in God, I do not conceive of saving Egypt from the Single Truth seen on television as delivering Egypt to some other orthodoxy. But one need not be totalitarian to recognize a single world-wide dominant cultural (or "social") structure, from which humanity and earth need to be liberated. Some language-games, like the buying and selling for profit game, rule the world. It is the dominance (not the existence—which is fine—but the dominance) of that basic buying-and-selling-for-profit game that made it impossible to continue (for example) the "Swedish Model"; impossible to continue because (for example) promoting social equality by raising wages (for example) tended to drive those buying and selling for profit overseas (Volvo to Brazil for example); and therefore the problem of reviving and deepening social democracy is (in part) the problem of making it more defensible than was Sweden's (for example) against the devastating juggernaut Bowles and Gintis call "the exit power of capital." If the dominance of one logic, one pattern of gainful activity, is undesirable because (among other reasons) it tends to make a nation defenseless against capital flight, then we will see Egypt's diverse activists more as promise and less as threat.

As I hear the familiar words of the dominant discourse I feel a tightening of breath; it is so coherent, so specious, it is repeated so often by so many well-dressed gentlemen and ladies; it is presented as the good sense of the good people, contrasted with the naiveté, the superstition, the fanatic violence, and the ignorance of the seething masses western civilization has not yet reached and drawn into its orbit of prosperity; I think of my mother in her dotage looking at her television set believing all of it; I think of the blank faces of thousands of students I have taught who come to college with no frame of reference for comprehending any of the alternative ideas the world at this juncture so desperately needs. But what I mostly think Is that I must stop watching television and start washing the dishes as fast as I can—since it was Miriam's turn to cook it is mine to wash—so I can get back to work explaining concepts and making constructive proposals, articulating a Game Plan for turning the tide.

I know it is crazy to feel that the young people of Cairo are depending specifically on me to explain the trap they are in and how to get out of it, since there is no reason to believe anybody there would be interested in the opinions of a retired professor down here scribbling away in a small rural town in the world's most southern and least probable country, but nevertheless, crazy as it is, that is how I do feel. I know the young people of Cairo could learn constructive alternatives to today's dominant discourse by reading many books already written, some by Karl Polanyi, by Hazel Henderson, Mfuniswela Bhengu, Evelin Lindner, Ronnie Lessem, Jose Luis Coraggio, Pierre Calame... some by me... but I still feel driven to keep writing. I feel so frustrated seeing my allies and I reaching the public so little, hearing the experts on and off television getting away with repeating over and over the same old arguments we have refuted a thousand times, while the people and the planet go on suffering and suffering and suffering; that I want to try something different, something new, something neither I nor anyone else has tried before, to get the message across. This book is different because I am taking a more personal approach, talking more about my life and my feelings, struggling to state my reasons so clearly nobody can pretend not to understand them, and at the same time sharing the dreams that drive the reasons, not because I know this approach will work but because in my desperation, in my frustration, I am ready to give anything a try. Although I believe that part of my motivation to compose these impassioned sentences is my present-day frustration in the face of constructive truth ignored and destructive error dominant; I also believe my crazy feeling that I am called upon to right the wrongs of Egypt, and the even crazier optimism that makes me feel I could, have roots in my youth and childhood, some of which I have already mentioned; and I have been told tales that make me believe my probably-somewhat-deluded-will-to-do-good grew from seeds planted during the nine months when I was an honored guest in the comfortable quarters of my mother's womb. Perhaps my mother had read my dear friend Plato, who did not specifically say that children should learn the good before they were born while still in the womb, but who did say they should learn good very early, so that when in later life they encountered evil they would reject the bad as an aberration from the world-as-it-should-be; or perhaps she had already garnered an inkling of the genetic epistemology then arising in far-away Geneva, which would discover how intelligence develops starting from interactions with the environment of behavioral schemas present at birth—behavioral schemas which themselves had grown gradually from interactions with an earlier environment before birth. My mother took me for rides before I was born, and after I was born my father took me for rides; she on the big red Pacific Electric trolley, Oak Knoll Line, to San Marino, where she carried me inside herself to the Henry Huntington Gardens and Art Gallery, where she stood before Gainsborough's "Blue Boy," Lawrence's "Pinkie," and other famous paintings there, as at home she would listen to Mozart on an old-fashioned gramophone, for the benefit of her as-yet-unborn first child whom she already knew would be a marvel; he, who so wanted to be a country gentleman, who subscribed to a magazine of that name, cruised the walnut groves of El Monte on his bicycle with his first and at-the-time only son riding in a box-seat above the rear wheel.

My transition from the-world-as-it-should-be to the-world-that-needs-me-to-fix-it happened between age five and age ten; starting in paradise on two islands of the world-as-it-should-be, one at my grandmother's rooming house next to a fig tree and full of cats in Pasadena, one at the half-acre in El Monte where my father dreamed of being a country gentleman; arriving in hell by age ten at a tract house in Fontana where my parents quarreled incessantly. Besides feeding me hamburgers soaked in melted butter, my grandmother would bloom ecstatic whenever I gave her pictures of cats, flowers, dogs, or grandmas labeled with one or two crayoned words. Most of her roomers followed her lead in praising the drawings she dubbed my "stories," being mostly like herself elderly unpublished poets with bad credit, including old Mr. Worrell, a blind gentleman, who wore a blue serge suit holding in its front pocket a golden watch on a golden chain, who pretended to be able to see them. "Little Alice" who lived in a small trailer parked behind the orange trees would reward me for my "stories" with oatmeal cookies made with raisins; and Harry Seaberg, who was born on a ship in mid-Atlantic half way between Stockholm and New York, taught me to waltz to the tune of Casey Would Dance with a Strawberry Blonde. My self-confidence and my communitarian idealism were equally nourished on our half-acre in El Monte, where the sheets on my mother's bed exhaled a super-natural softness; where my father would take down a peach from one of our trees and cut it in half with his jackknife, half a peach for him, half a peach for me; where I was supposed to tether the goat but the goat ran away dragging me along the ground as I faithfully held on to the rope; where I first learned the names of the vegetables and of the flowers; where the ducks cackled with delight when they were turned loose in the beds of iris to eat the snails --Pasadena in 1942, El Monte in 1944, two places where I lived; where I still live today because, as Uncle Sigmund has taught us, all the times of our lives are simultaneously present to us in our subconscious minds; where I drank into the depths of my soul an optimism that does not die, that did not die when wartime prosperity faded; that did not die when my mother divorced my father; that did not die when my father went insane; that did not die when in Frederick Jameson's words "the sixties ended on September 11, 1973, in Santiago, Chile," with me as an eyewitness, watching from the window of a fifth floor apartment on Cummings Street as the Chilean Air Force bombed the presidential palace where its commander-in-chief, Salvador Allende, was preparing to shoot himself; an optimism that survives to this day as a deep blue calm under everything. By the age of ten (1948) I felt my family counted on me to save it not just because only I could possibly reconcile mother and father, but also because of the four of us I was the only one with an income, bringing a bit of cash into the family coffers by selling newspapers on the streets after school at five cents a copy, of which two and one half cents went to the publisher, while the product of the remaining two and one half cents multiplied by the number of newspapers I sold went to my family. My father was a truck mechanic who had been steadily employed during World War II, but as normality returned and the supply of workers once again exceeded the demand for workers, he found himself permanently located near the bottom of the barrel, among those would-be sellers of labor-power with relatively outdated skills and relatively unpleasant personalities who never made the short list, but my mother was not allowed to work because my father insisted that he had to be the family bread-winner. We survived on charity and on subsidies from grandparents supplemented by my meager earnings as a paper boy until our mother left our father and took us (my younger brother and me) to Barstow, a railroad junction on the desert where nobody who had a choice really wanted to live; where the School Board was so desperate to staff the classrooms that it hired my mother to teach third grade after she took (on a scholarship) one summer school course on how-to-teach, provided that she spend her subsequent summers studying for a teaching credential, and paid her —as a trainee studying to become a teacher—two hundred and twenty five dollars a month. Faced with the question how to feed herself and two hungry boys on her salary, my mother gave up and assigned to me the task of calculating the answer. My answer was potatoes, rye bread, cabbage, and two eggs each per week. Apples I bought too but instead of eating apples we traded with Mexican neighbors for flour tortillas, because the tortilla was then our optimal choice to maximize calories per dollar and we did not know how to make tortillas ourselves. When a freight train wrecked dumping huge quantities of yellow-green grapes beside the tracks, we and other poor locals gleaned them and ate fresh grapes until they got too old; and with the help of our friend Alma—another teacher trainee—we dried some for raisins and preserved some. I took care of my kid brother during the summers when our mother studied for her credential and we were shipped off to a foster home in the San Bernardino Mountains near Mount San Gregorio, a mountain affectionately known as "Old Grayback." I did a pretty decent job taking care of my brother and mother; carrying my brother down the mountain when he fell ill as we were scaling "Old Grayback;" decorating my mother's "environment" (her classroom); but I was scared, scared of being beaten up by other kids, scared of our father's revenge, scared of our father's insanity. Our father formed the habit of sleeping late, breakfasting in his bathrobe on coffee and doughnuts in his bedroom at the boarding house, then closeting himself there ruminating about superannuated grievances; donning shirt and pants around three; taking supper at six, overeating, sleeping until midnight; then pacing the living room until dawn, walking around it smoking in the dark—one could trace his elliptical orbit watching the orange glow of the burning tip of his cigarette.

He spoke a private language only his mother and his two sons could understand, composed mainly of disguised complaints about his unemployment and his wife deserting him, and of futile attempts to exercise an authority he no longer had, like, for example, "My name is Kenneth and I am an American," which meant immigrants got jobs while he did not, and "Where's your mother?" which meant the divorce was illegal and his marriage still existed. When my father told me that the Nazis won World War II and put the German general Dwight Eisenhower in charge of America I could never tell whether he was complaining via extended metaphor about my mother leaving him and putting me in charge of the family, or was playing his role still being my father in spite of the divorce by telling me facts I needed to know. My father's incompetence fueled a fear I might turn gay that drove my mother to beg the school principals to assign me men teachers to model manhood, to inspire me to copy men, induce me to desire women. The men teachers—Mr. Schumann, Mr. Cork, Mr. Bromberger, Mr. Leinkamper—were four stalwarts in their twenties who had been to Europe and fought there, who had been rewarded for their World War Two military service by a college education at the government's expense, who were the first in their families to realize the dream of practicing a profession albeit the lowly profession of teaching school, who backed Franklin Roosevelt's progressive politics; who saw themselves as the intellectual elite of a crumby little town, and who eagerly assumed the role of mentors to a bright little boy with a pretty divorcée mother—but who were quite different from my previous intellectual mentors, the impecunious elderly ladies who would form a circle of chairs in the front room of the boarding house (or of the foster home) in the evenings to talk about God and family; for example a retired cleaning lady from Boston who rented a room from my grandmother who gave me a Bible I still have autographed, "For Howard in the holy name of Jesus, from Nola Bartel."

By age twelve my young male mentors and my elderly female mentors had accustomed me to thinking in terms of the Big Picture. I understood that my father's unemployment, sexism, racism, and insanity posed cosmic issues. My formal commitment to being a philosophical do-gooder came at age fourteen as a result of night terrors, panics that according to the leading expert Harry Stack Sullivan you cannot possibly understand if you have not experienced them yourself; so unless you have had your own night terrors you will probably not understand how I escaped mine by making a deal with God that was actually not a deal since He (She?) did all the talking—theologically it was a covenant, not a contract; a meeting of minds of unequals, not of equals—and what God said was that in this world so driven by violence, fear, anger, sloth, indifference, economic calculations and lust my job was to use reason to try to devise a functional order; which was a tough but not bad assignment, and in any case I was in no position to bargain since I would have done anything to banish my night terrors.

One might say my work to articulate a Game Plan to transform the modern world-system that traps the young people of Cairo is tainted by false motives since underneath I just want a job for my father, to cure his insanity, to unite my family, to escape my night terrors. But one might also say that anyone so strongly motivated to seek solutions might well have found some solutions, and if one took this latter tack one might be willing to consider a fundamental and far-reaching principle this particular seeker has found by research and reflection: human beings are not naturally predictable like the planets in their orbits; we make ourselves predictable by organizing ourselves. The organization of human life is about ethics, about human action, not about physics, not about mechanical action; it is about inventing and following rules; so when a culture is

dysfunctional it is either because of *anomie* (disor-
ganization) and/or because today's organizing rules
themselves cause dysfunctional behavior. Caveat: rules
only work when soaked in myths, dreams, ceremonies,
good relationships, old habits. Caveat: nature judges
culture in the end; we cannot simply invent cultures
that ignore ecology and behavioral biology.

As l develop this idea of social science as ethics
not physics by relating it to some ideas of Paulo Freire,
Max Weber, Karl Marx, the Grenoble School (Michel
Aglietta, Robert Boyer...), Hazel Henderson, John
Maynard Keynes, Karl Popper; and to one of my own
ideas, "cultural resources," I think you will see why I
think this is stuff the young people of Cairo urgently
need to know about, and I think you will at least begin
to see how a Game Plan for turning the tide is born
from my conceptual matrix. I think you will see that
the themes of "inventing and following rules," "basic
constitutive rules," and "basic cultural structures of the
modern world" provide lenses for understanding both
the contemporary global economy and the classics of
the social sciences. Think of Paulo Freire´s landless
peasants in North East Brazil, who—while landless—do
have Adam Smith´s natural liberty, natural liberty that
Charles Taylor calls constitutive rules of a bargaining society, that I call basic cultural structures of the
modern world. They are free to buy what they have
money to pay for. Peasants and landlords are formal
equals in the market place, and the agreements to
exchange the property of a peasant (labor-power) for

the property of a landlord (money paid as wages) are contracts where the peasants of their own free will agree to the terms, since they are free to sell their labor-power or to refuse to sell it. In his article, "Cultural Action for Freedom," Freire outlines a method for freeing the peasants, but the peasants are already free according to the dominant ideology. Because Freire wants to change contract and property rights, that ideology counts Freire as an enemy of freedom. Issue is joined on the question: what does it mean to say "freedom?" Freire says "freedom" is something the landless peasants do not yet have, which is to be won by acting to change the culture's basic rules.

Prior to cultural action, prior to the peasants enrolling in Freire's "culture circles" and having their consciousness raised, they regard their fate as natural because the social order has been in Freire's terms "mythicized' to make the peasants think being oppressed is natural. The myth that makes the peasants see their oppression as natural is the myth that calls them juridical subjects, already free: it is the myth of natural liberty. What the peasants see when their consciousness rises is that their oppression is cultural; it flows from myths portraying landlords with land and peasants without land as the way the world has always been and will always be, it flows from today's basic constitutive rules; oppression does not flow from natural laws nor from human nature; just as similarly feminists using Freirian methods facilitate women (and men) coming to see that patriarchy is not natural but cultural; just as, conversely, we can describe a social science as consciousness-lowering when it treats Smith's version of liberty and justice as natural. Max Weber tempers denunciations of oppression-under-law with nuanced praise for formal freedom, reporting that in early modern times serfs on Junker-ruled estates in eastern parts of Germany would flee the cruel tyranny of their masters to breathe the free air of the cities, even though in a city they were no richer, and often poorer.

More importantly for Weber in a capitalist world nobody is free since the system is an irresistible economic "machine," an "iron cage," and all who are trapped in the cage must obey the machine. If we look carefully at what Weber means by the metaphors of the "machine" and the "iron cage" we are led back once again to the basic cultural structures of the modern world, to the box, to Smith's "natural liberty," to Marx's "Eden of the innate rights of man," to the basic constitutive rules; beyond Weber's famous thesis that an inner-worldly puritanical asceticism motivated the early capitalists to accumulate (instead of spending their profits on luxuries), to something Weber says even more clearly: that accumulation is made possible by a neo-Roman legal system that protects property and enforces contracts, thus constituting an iron cage made of normative (legal/ethical) bars, thus constituting the rules that make the logic of accumulation possible, and often apparently necessary, for it often seems we have no choice but to play the game of profit-making, even if we play that game only as employees of profitmakers; and so we see that "machine" is a metaphor that when cashed out to refer to its literal referents proves to be an "iron cage" made of games with rules.

Since rules (or norms) are regularities in conduct backed by authority (Durkheim); since rules are licenses to criticize (and sometimes punish) those who violate them, and guides people use to monitor and direct their own conduct (H.L.A. Hart); since rules are organizing principles that constitute institutions; human life without rules would be physically impossible. Humanity's key problem is that the particular set of basic rules now dominating both daily life and the global economy nails us into a box, traps us in an iron cage; our key solution is to change the rules. Let us see how rules constitute the "economic machine" analyzing a diagram like one Karl Marx uses in the second volume of *Capital*. The "machine" works this way:

$$M - C --- P --- C' - M'$$

The diagram flows from left to right, starting with M, money, then moving from M to C, commodities. The sequence M – C means first capitalists buy labor-power, equipment, raw materials, and generally everything needed to produce something, all of which are here called C, "commodities," i..e. "things you buy," also known as merchandise; Marx calls them *Waren*, a German cognate of the English "wares" which Simple Simon wanted to taste, but which the Pieman would not let him taste until he paid. Next in Marx's diagram the Commodities the capitalist purchased are put to use in Production, P, in which the labor-power uses the equipment to process the raw material to produce another commodity, this time C'—greater in value than the original C that the capitalist purchased; C' is then sold to yield M', a sum of money greater than the sum of money originally advanced. The net result is to transform money into more money. The output M' (the augmented money) can then be cycled back as input. The new input M' becomes M'', a still larger sum. Production depends on money accumulating forever. The key to understanding accumulation (and to changing it) is to see that each step depends on rules—on basic cultural structures. Accumulation begins with purchases of commodities, that is to say with contracts because a purchase is a contract, and then continues with the exercise by capitalists of property rights over the use of labor power and other items purchased. Roman jurists debated who owned the product C', but now the legal rules make the owner of the business undoubtedly the owner too of the increase in value resulting from production. Even more fatefully—and you have to understand this if you are going to understand anything at all about the modern world—the *motivation* that gets production started—that determines whether there will be employment or unemployment—is that increase in value. Note that the sales that turn value added by production into cash are governed by the legal rules of contract. The idea of "accumulation," the repeated cycle of money turning into more money, can be extended—as the Grenoble School has extended it—to show that nobody is free, not even the more than six billion humans on this planet acting collectively —assuming we could act collectively—would be free, because we are all compelled to obey the systemic imperatives of "regimes of accumulation." Here is Grenoble (Michel Aglietta, Robert Boyer ...) in simple terms: "Look guys: If nobody advances money (M) to get production (P) started, then there ain't nothin' to buy (no C'), there ain't no jobs, there ain't nothin', but nobody advances a dime without thinkin' there's somethin' in it for me, namely getting back more than a dime, so guys, the one thin' we got to do, before we do anythin' else, because if we don't do it we can't do anythin' else, is to set up everythin' to boost profits, to keep them profits rollin' in." Schools, police, sports, music, politics, psychology, culture, science, television, family, ... everything. "Regime of accumulation" means all rules must favor profit. If one "regime" falls, as "the age of Keynes" fell, another "regime" forms, like "neoliberalism."

Hazel Henderson represents today's growing trend to thinking outside the box. And trend to *seeing* outside the box. We have both incentives and opportunities to design ethical futures free of the imperatives of the economic machine, and even now a majority of the world's work is not done in the "machine" described by Smith, Weber and Marx; for examples: the people's economy where the objective is to eke out a living in a tiny business that makes no profits, parents caring for children without pay; the public sector; cooperatives; non-profits; subsistence farming, do-it-yourself home improvement.... Henderson calls for acting ethically, offering tons of upbeat ideas for "managing socio-economics," and she is obviously not just talking to a brick wall because many—including some major corporate executives—respond to her; they put her on boards of directors, they set up ethical funds to be ethically managed, they rethink missions and visions. Today's trend to ethics follows yesterday's failed consensus. Not long ago there was a near-consensus among sensible people that thanks to Keynes the way had been found to run mixed economies; and thanks to Karl Popper everyone sensible favored an open society.

The party had to end because, given the paradigm in place, Keynesian policies of government spending and central bank lending could only lead to stagflation and mountains of unpayable debt.

That is why I am writing a short book about everything, addressed to educated people of good will who see the need for paradigmatic change. Popper's open society could not be realized because, given the box, political power could not dominate economic power. That is why we need a philosophy modifying Popper's articulating a new synthesis of the social and natural sciences that provides an overall framework for deciding what to do.

I believe that the young people of Cairo, now that they have overthrown their dictator, should be discouraged from believing that Europe's yesterday can be their tomorrow; and encouraged to believe that the religious and political groups television portrays as threats to investor confidence might bear the seeds of constructive alternatives. Of course I may believe this just because I am a little bit crazy and it may be my mother's fault. But it just might be the case that my mother did the world a favor by bringing me up to feel that it was my appointed task to figure out solutions to its problems, and then to keep pestering it to get its attention. It just might be the case that the young people of Cairo need to know that while liberal logic reigns the exit power of capital makes Egypt defenseless against capital flight, that the particular basic rules now dominating both daily life and the global economy nail us into a box, that today we have both incentives and opportunities to build ethical futures free of the imperatives of the economic machine.

Four

"*Ubuntu*" describes a magical childhood I never had and still seek. It is an old Bantu word; for many Africans it is a treasure to preserve; it names the human values of pre-colonial culture. Mfuniselwa Bhengu, who heads a Centre for Economic Humanism and who represents Cape Town in South Africa's parliament, has written about *ubuntu*. In three books he has summarized its history and philosophy. Reading in Bhengu's books the missing pages of my life, I learn that in the times of *ubuntu* and still today where *ubuntu* is practiced there are no old age homes and no lonely seniors because the members of each extended family care for their own; for the same reason there are no orphans. "There are always arms outstretched waiting to receive relatives in distress; foster parents who eagerly snatch up orphaned children;" brothers and sisters galore to rally around the ailing or crippled worker; equally as many available to provide granny and grandpa with food, clothing, and shelter; big crowds at weddings because everyone in the community is invited—but they do not come empty handed but with beer and food as well as gifts for the bridal couple, and they join in the dancing and singing; while a widow, according to custom, moves in with her deceased husband's eldest brother; and thus *ubuntu* is experienced daily as a living reality by children, by teenagers, and by adults. Such marvels of convivial practice are produced by myths—like conceiving the family as consisting of the ancestors, the living, and the unborn—and indeed by a whole cultural matrix, for *ubuntu* is an ancient universal philosophy, a collective respect for human dignity, the potential for being human, the art of being human, valuing the good of the community above self-interest, being honest and trustworthy, fairness to all, compassion, the element of godliness in a human being, the spiritual foundation of African societies. *Ubuntu* begins when grandparents participate in birthing, and are first to hold the newborn, because since she or he is considered a villager who has just arrived from a journey that started in the land of the ancestors, the newborn is most at home with the elderly. No one is born on this earth without a reason, a special purpose, and the old help the young to remember and claim their purpose in life in a supportive village atmosphere based on trust where no one has to hide anything. Mfuniselwa Bhenghu writes: "Throughout children's life in the village there is a strong message that they belong to a community of people who value them almost beyond anything else." "Community grows in an atmosphere in which people can drop their masks." If you are a

Karl Polanyi or a Marcel Mauss, you will take this singing of praise for *ubuntu*; and similar praise that could be sung for ways of life of indigenous peoples of the Americas, Asia, early Europe, the early Middle East, Australia, and the Pacific Islands; as evidence that the extreme individualism of the cultural structures of the global economy is an aberration in the history of the species, which tends to destroy even its own conditions of possibility; and as evidence that in the future, if humanity has a future, humanity will regress toward its norm, rediscovering *ubuntu*. Agreeing with Polanyi and Mauss, I take *ubuntu* in particular and community (Toennies' *Gemeinschaft*) in general to be the key missing ingredients in social democracy, for lack of which social democracy fell and neoliberalism rose.

When I search my memories to find—if I can—my own experiences in daily life of *ubuntu*—or something like it—my thoughts drift toward watermelon and softball under a tall eucalyptus tree with crowds of Minardis young and old, the extended Italian-American family of Guido and Margaret, my foster parents for a summer when my mother did not send me to the mountain home but to a home in a semi-rural suburb of Los Angeles; where Guido lined us kids up every morning to assign the work for the day before driving to his own work in the city; where I learned to be a rough carpenter nailing down the floorboards of the bigger house the Minardis were building on the same lot. Margaret hugged me more than my real mother did; she told me of her earlier life at Disney Studios making drawings for "Snow White;" she shared her Catholic faith; she once put a bandage on my hammered finger that I kept there for months, gluing it back whenever it fell off until at last it wore to shreds.

I feel again the *ubuntu* spirit when I visit the Rosenkranzes in Buenos Aires, a city that can give the impression of being inhabited by hordes of strangers, individuals living in concrete buildings along asphalt streets in environments bearing no resemblance to the grasslands and wooded hills where the human body evolved, as the commercial form of their social interactions bears no resemblance to a clan or a tribe. But inside their fourth floor flat are: thirteen grandchildren doing a sleepover with grandma and grandpa; one heart-shaped homemade sign declaring Abram to be the world's best grandpa; thirteen small heart signs each inscribed with the name of a grandchild; a dual calendar telling what day it is both by modern reckoning and by the ancient Jewish reckoning. Friday evenings the table holds a full glass of Kosher wine blessed by skull-capped Abram chanting instructions for keeping the Sabbath holy in Hebrew, a language he and Sara learned in the Hebrew School of this neighborhood where they were born, where they live, where they expect to die; studying alongside the other little boys wanting to turn thirteen and be *bar mitzvah* ("son of the commandments") and the little girls wanting to be twelve and *bat mitzvah* ("daughter of the commandments.") Sara and Abram are full-fledged moderns, she a distinguished professor at the University of Buenos

Aires, he the owner of a small textile business; and yet they are still members of a tribe, of a close-knit family in a close-knit community where it is impossible to be destitute, impossible to be an abandoned child, impossible to be abandoned in old age. I offer the Minardis and the Rosenkranzes as evidence for Mfuniselwa Bhengu's thesis that *ubuntu* can be read as the global philosophy for humankind, as an acceptable proposal for defining what it means to be human (in the Bantu languages *ubuntu* means "human"); applicable everywhere while yielding Africa pride of place as the site of the origins of humanity where the first ancestors of all of us lived no matter where we their distant grandchildren may be living now; as the place where our bodies and feelings evolved; where our deep emotions were programmed into our souls. My heartfelt question about *ubuntu* is "How can I get more of it?" because I feel I had too little; because my family is too small; because the bonds that bind it are too loose; because I fear I may spend my old age like my mother, alone in a hospital bed with only one regular visitor; but beyond bemoaning my personal fate, which is the fate of millions, billions, I realize my old age insecurity –like my father's unemployment, his sexism, his racism, his insanity—poses cosmic issues; the disenchantment of the world and the rise of individualism are phenomena whose causes must be understood by historical study and whose cures require a Game Plan. Now I will proceed to work on the historical study of the decline of *ubuntu* and on a methodology to cure disenchantment; hoping that readers will not commit the fallacy of assuming that because I long for the security of a world where the members of extended families care for one another, therefore I must believe modernity to be all-bad and pre-modernity all-good; hoping to hinge Freire-style the charms of *ubuntu* to green paths of life, to *los caminos verdes*, to sustainability for our human species as well as for our plant and animal companions so dependent on us to preserve their and our habitat here on this small blue planet tucked away in a remote little corner of the Milky Way.

In his 1893 doctoral dissertation at the University of Paris Emile Durkheim offers a causal analysis of the decline of the traditional societies organized by clan and tribe; that is to say, organized as was normal for the human species for hundreds of thousands of years; organized by sanctifying reciprocal obligations of mutual aid within extended families (he calls them *societés segmentées*). While population densities remain low, Durkheim theorizes, *societés segmentées* can meet their material needs, but when populations increase kinship-organized societies inevitably decline. Supporting a large population requires what is now called "development," which in Durkheim's time was called "division of labor," or alternatively "civilization;" "division" because millions of workers each one dedicated to a specialized task exchange goods through markets; "civilization" because markets require law and order. Durkheim echoes Adam Smith who wrote a century earlier, for the starting point of Smith's analysis is the size of the annual fund possessed by a society to supply its wants and conveniences during a year, while the main cause that determines the size of that fund is the division of labor. Book One of Smith's *Wealth of Nations* contains an argument explaining why the people Smith identifies as "savages" are poor, while those he calls "civilized"—primarily his fellow citizens in Scotland and England in the 18th century—are rich, making the claim that the "savage" is a jack-of-all-trades-master-of-none who tries to do everything for himself, while civilization dawns with division of labor, in which each does what he (I do not correct Smith's pronouns because I believe his meaning as well as his grammar is sexist) does best; at first bartering with neighbors to supply his needs and later—more efficiently—selling his products for money and buying the goods to supply his needs with money; and because when everybody develops a specialty in what they do best production is incomparably greater it follows, says Smith, that even the common workman among the civilized enjoys incomparably more of the wants and conveniences of life than even the king among the savages, because the cause of wealth is division of labor, while the institution that makes possible the division of labor is the exchange of products in the market, while the legal rules that make possible the functioning of markets are the rules of property (making each the owner of what he brings to market) and the rules of contract (organizing buying and selling); the very same rules that constitute what Smith calls "natural liberty," whose observance is, in Smith's terminology, "civilization." Reading the rest of The *Wealth of Nations* we learn that Smith overstates his case in Book One when he makes the division of labor almost the only cause of wealth, for he goes on to give equal weight to the accumulation of capital, for—he says—without a capitalist advancing the worker his means of subsistence (in the form of wages) during the time when the product is not yet ready to be sold, there can be no production; nor can there be investments in what Smith calls "improvements" of productive techniques without accumulation of capital. Smith and his friends can explain the decline of *ubuntu* values as a consequence of learning that emerging from poverty requires less caring and sharing and more accumulation, less warm personal bonding and more cold commercial transactions; but Walter Rodney and his friends explain history differently: Europe seized what Rodney calls a "temporary military advantage" to conquer and to colonize Africa and to impose on it educational systems and religions designed to favor European interests. If with Rodney we decide to attribute some major part of the decline of good traditional values, in Africa at least, to the military conquest of one culture by another, then to complete the causal analysis we should explain the military superiority of Europe, and I know no better explanation than the one given by Adam Smith in Book Five of The *Wealth of Nations*: the accumulation of capital; for the same capital that when invested in industry produces wealth, when invested in fleets and armies produces power; so therefore, says Smith, the nations he calls "opulent" overawe the poor nations. For Marx, since a culture grows from its material base, it withers and dies when cheap goods from the modern sectors make its mode of production uncompetitive. For me, the basic cause is the basic modern cultural structure. Our basic cultural structure, our civil law, our normative framework for the division of labor, is what Durkheim analyzes as a demographic necessity and names "organic solidarity;" it is what Adam Smith calls "natural liberty." It is the legal and moral framework of a market economy, and therefore of the accumulation of capital that made possible what Rodney calls "Europe's temporary military advantage" and the violent suppression of *ubuntu*. For Marx the basic cultural structure of the modern world is what he mocks as the "Eden of the rights of man" that frames buying and selling, and therefore frames the competition where traditional craft-based societies lose out to modern industry. The good news is that cultural action is a realistic methodology for change because insofar as cultural structures have causal force in history—I do not say they cause everything, or that they are one hundred percent of the cause of anything—cultural creativity is a force that can change history. Since it cannot be too often repeated that the main goal of cultural action is to change cultural structures, to change the rules—in Paulo Freire's terms to humanize both oppressors and the oppressed—and not just to change who has power (important as that is) I will again repeat the point that the systemic imperatives of the modern world-system are constraints constraining solutions to any problem whatever, this time taking my text not from Adam Smith or from Karl Marx but from John Maynard Keynes.

Reading Keynes' *General Theory* drives home two crucial points: first to the extent we are still in the box, still in the iron cage, we must do what the system commands, even when that means not doing what ethics and ecology command; second, exclusion: while we are in the box we have a class of people like my father, would-be sellers of labor-power who find no buyers, tending toward crime, alcohol, drugs, and as in his case mental illness. Reading Keynes helps us appreciate *ubuntu* as an emancipatory cultural resource, for *ubuntu* frees us from the four sides of the iron cage: service to others replacing Bentham's principle of self-interest, sharing and stewardship changing the meaning of property, culturally determined behavior complementing contracts, responsibility reconstructing freedom. To discussions of what Max Weber calls the "iron cage" of the "economic machine," Keynes adds a discussion of "confidence," starting by saying it is not fully accurate to call the "machine" profit-driven. It would be better to say expectations of profit drive investments, but Keynes does not believe anyone can rationally calculate the expectations of profit ten years hence from a railway, a copper mine, a patent medicine, or a sea-going vessel; and in much of practice profits are made not by predicting correctly what will happen in ten years, but by forecasting the short term psychology of the stock market, where shares in railways, mines, pharmaceuticals, and shipping are traded daily; the object being to predict the state of confidence of thousands of investors all of whom are, like oneself, also trying to divine the state of confidence of everyone else—a process Keynes likens to a contest in a British newspaper for picking among one hundred photographs the six prettiest faces, with a prize for the entrant whose picks most closely match the average picks, where the objective is not to choose the six really prettiest, nor to give one's sincere opinions, but to guess which six faces everyone else will judge prettiest.

It is absurd but it is real, for the welfare of billions really does depend on gamblers predicting how others will gamble.

Hopefully Keynes' analysis of its absurdity will help more people learn to think outside the box. Inside the box one says, "Egypt has a lack-of-confidence problem , so therefore measures must be taken to give investors confidence in Egypt," while outside the box one adds, "True, at this point we have no choice but to comply with the systemic imperatives of the iron cage, but we can also think creatively about making livelihoods less dependent on the ups and downs of stock markets," and if we make this outside-the-box addition to our thinking we will then search for the cultural resources that are even now mobilizing resources to meet needs, giving people some measure of security and happiness *malgré tout*, and if our eyes are open to see and our ears open to hear, we will see them and hear them, for the spirit of *ubuntu* is not dead; it lives on amid the concrete canyons and ringing cash registers of modernity, and it can be enhanced.

Take ABCD, the Asset Based Community Development movement that began in a black church in an American "ghetto" or "inner city" --two synonyms for "place where nobody invests"—when the members realized they would have to revive the neighborhood or else move away. Unable to count on private investors, on government grants, or on non-profit organizations, the members of the congregation asked, "Who can we count on?"; answering "Ourselves!" then proceeding to list what skills, experience, property, and knowledge each could contribute to a group effort to turn the neighborhood around; thus inventing the germ of a method for "mapping" a community's resources and so jumpstarting using the resources to meet people's needs. In the version of "mapping" I have used working with Catholic groups in California we call it "inventory of gifts," listing our "gifts of the heart" (enjoying teaching children, caring for the old, tending gardens ...), "gifts of the hands" (auto repair, carpentry ...), "gifts of the head" (accounting, computers ...), "gifts of property" (a building, an orchard, a pickup truck ...) etc. Each gift of each person is noted on a separate card; then the cards are scotch-taped on a wall. Then we move them around, taping them together in different combinations, seeing different ways to combine gifts to start projects.

The cultural resources of Sri Lanka's *Sarvodaya Shramadana* movement are steeped in ancient wisdom and local customs, its Sanskrit name meaning literally "the awakening of all through the sharing of labor"—a phrase that exactly describes how its members begin to "awaken" a village steeped in apathy: identifying a need, such as latrines, a school building, a well, or a road, and then donating labor to meet the need; always building on—never deprecating—the value-system the villagers already have. "*Sarvodaya Shramadana*" does not at first sound like "*ubuntu*" since one speaks Sanskrit languages, one speaks Bantu languages, but closer inspection shows they share common features: mutual aid among bonded groups, respect for author- ity, prizing being more than having. Today these old- fashioned ideals serve new key goals: they free us from the iron cage (Weber), from systemic imperatives (Ellen Wood), from the necessity of a regime of accumulation (Grenoble School), from dependence on the state of confidence in stock markets (Keynes). When we depend less on the system we are more free. But although *Sarvodaya* and ABCD are clear small examples of norms which loosen the iron bars that now constrain solutions to any problem whatever, to turn the tide we must enlist massive movements, such as corporate social responsibility, micro-financing, and social safety nets run by governments. For example: the corporate social responsibility of Mexico-based CEMEX, the world's third largest manufacturer of cement (which happens to have four plants in Egypt). CEMEX takes building social capital as a goal, thus reversing the individualism that started modernity. I want to write clearly because this stuff is very important. CEMEX's desire to be responsible can surprise only economists; everyone else knows humans want meaningful lives.

In the Mexican recession of 1994-5 CEMEX's sales to high end customers fell 50 percent, while sales to its poorest customers fell only 15 percent, illustrating that the carriage trade is more volatile in the global casino than selling to the poor. In Latin America and Africa the majority ekes out a living outside the profit economy, not making profits and not working for someone who does. When for economic and social reasons CEMEX chose to concentrate on selling cement to people who could not show a balance sheet or a pay stub, it quickly learned that they wanted not precisely cement, but to build a do-it-yourself home one room at a time—that was their dream—and they already had an indigenous form of pooling resources to fulfill it known as a *tanda*; and so CEMEX partnered with the people's economy, organizing self-help clubs, giving architectural advice, teaching classes on masonry, selling cement and related items on credit not to individuals but to three-person groups. Thousands of homeless now have homes; thousands of families who lived in one room now live in two; thousands of idle youth now have real work to do building an addition to the family home; most importantly: the individualizing juggernaut of neoliberalism has been reversed: people are more connected, more bonded, more secure.

I wish the micro-credit schemes for micro-enterprises that today mushroom everywhere, had existed back when I was ten years old to provide my father with something to break his fall into permanent unemployment. My father could have been like the street vendor José whom I greet whenever I take a stroll downtown, always standing on the same street corner, always having a kind word for each of the passers-by, especially for his steady customers like me, offering a choice of a small made-in-China pack of vitamin C pills, or band-aids, or (in season) Christmas gift labels, all at one hundred pesos; usually standing not far from another micro-entrepreneur, Eva, sitting on the sidewalk near José with a white plastic-foam thermal box containing *humitas*, a corn dish similar to the Mexican *tamale*, for sale at three for a thousand pesos; my father could have gone to night school courses for micro entrepreneurs like the ones at the training center on Serrano Street; he could have gone to meetings with other micro entrepreneurs like the ones at the Soul of Limache Club to exchange ideas, sing songs, and listen to motivational speeches -- not making much money but having something to do and making enough to survive if (as in Chile) health care and education for the poor are free and housing almost free.

If you ask why social safety nets for José and Eva are growing not unraveling, the answer is copper: the government owns the old mines and taxes the new mines -- a point of theoretical importance highlighting Ricardo's thesis that taxes on natural resources can fund public expenses without slowing business activity or causing unemployment. I fear I appear to cheer a mere seven-tier career to get to a plural economy from here: grassroots self-reliance, love, responsible big business, people's economy, do-it-yourself, micro-enterprise, and public safety net. My real point is broader: when you think outside the box you can see many ways to mobilize resources to meet needs: not seven but an unlimited number. A key goal of a plural economy is freedom from the imperative to curry market confidence at any cost.

As I see the world, the solutions to its principal problems are easy to invent but hard to explain. As I see the causes of historic events, the eclipse of the spirit of *ubuntu* was caused by the installation worldwide of the impersonal rules of a market economy and with them the imperative to establish investor confidence and whatever else the system may require, even when doing what the system requires conflicts with what humans and the biosphere require. As I see the Game Plan, whatever were the causes of the causes (the causes of the installation of the present basic rules), the priority now is to loosen the constraints that make it impossible for humans to govern the system, and the way to achieve that priority goal is to develop cultural resources to build a plural economy where no single systemic dynamic dominates the power of the people.

There are infinitely many such cultural resources. They are the organizational capacities for meeting human needs. Let me give another example to emphasize the point. Aravind Eye Care does 190,000 eye surgeries a year. Founded by the famous "Doctor V" (Venkaraswamy) it is organized as a nonprofit foundation under the laws of India. Although Aravind has created very efficient low cost techniques, even so the majority of its patients cannot and do not pay—but their bills are paid neither by government nor by raising funds soliciting contributions from donors. Driven not by profit but by the ideals of Dr. V.—carried on today by the staff—Aravind takes in enough money in fees from patients who can pay to cover serving patients who cannot pay. The people who work in the organization inspired by Dr. V's vision are biologically members of the same species as people who think and feel the mythology of Adam Smith in their very bones, but they live by different myths, driven by different dreams. Aravind is a proof that *ubuntu* is alive in our twenty first century; humans can still be human. The Game Plan we need will locate and nurture growth points like Aravind; it will showcase signposts like management professor C.K. Prahalad's study of Aravind that bring cultural creativity into mainstream thinking; it will outline what you and I can do to help all humanity —now more the object driven by the global economy than the subject designing it—to escape from what Max Weber called its "iron cage." Let me now give an example showing how the escape can be engineered.

When a Chilean chain of hypermarkets named Jumbo offered to construct a huge retail complex in Rosario, Argentina creating jobs for 2,500 people, asking in return permanent exemption from all municipal taxes, the City Council of Rosario was not afraid to reject Jumbo's proposal because—here is my point— Rosario already had in place multiple ways to create livelihoods for its citizens. Rosario has an office of Economic Solidarity backing micro-enterprises. It has urban agriculture and employee-owned businesses. The Argentine national government pays for employing the jobless in community service.

This example from Rosario shows how *ubuntu* just might make a comeback. What destroyed it might be overcome. As for instance Rosario's plural economy grounded in ethical solidarity provides leverage in negotiating with multinational capital, in general outside-the-box thinking motivated by human and ecological values loosens the iron grip of systemic imperatives; pre-history ends; the members of the human species become less passive victims of history and more active agents able to shape their personal lives and their collective futures. Now we are no longer in the boxed world where everything old-fashioned (extended family, customary forms of cooperation and sharing, spiritual traditions...) and everything futuristic (utopian visions, ecological design, democratic social-ism ...) had to be sacrificed (in the name of science no less!) to alienation calling itself progress. The laws of economics follow from certain social norms. Since norms can be changed, it is possible to do things that according to neoliberal economics are impossible, such as reducing today's extreme levels of social and economic inequality. And such as sustained full employment, which is an outcome not in the cards in the box; it is not an outcome that is going to happen while we play the game we are now playing—but it is a possible outcome.

It is possible complementing the employment the system delivers at a given time with additional livelihoods governed by other social norms. To the extent that we organize livelihoods by *ubuntu*, or by any alternative logic, we make the impossible possible. Bolivia is a country that provides new models; for example spending natural gas export proceeds on appropriate technologies supporting ancient *quichua* and *aymara* ways of life.

Of course less unusual models, like cooperatives, nonprofits and public sector employment, also contribute to freeing us from excessive dependence on investor confidence; they too help us get past the systemic imperative that requires that one group of people expect profits before the people in other groups can get jobs. And of course escape from having to obey a single capitalist logic is merely emancipation. It merely clears the way for building cultures where there are always arms outstretched to receive relatives in distress, foster parents who eagerly snatch up orphaned children, brothers and sisters galore to rally around the ailing worker. More than freedom is needed to achieve in time future the magical childhood I and many sorrowfully lacked and to prevent in time future the lonely old age I and many desperately fear, and—speaking generally—to achieve Durkheimian unity governed by norms; something like *ubuntu* is needed, a collective respect for human dignity, an art of being human, an element of godliness, a spiritual foundation.

Five

Idreamed afterward of flying when we made love together the first time. I soared among clouds teasing me like feathers; taking my breath away over and over everywhere and nowhere; as I had thrilled her and she me. Still today I live in my dream after our first tryst, reminded of it daily by the little finger of her left hand, by the lower lobe of her right ear, by her fourth left toe. And the tip of her tongue, the music of her voice, the heat of her breath. But let me tell you also about another dream I remember; a dream of which I am also reminded daily; a dream to be interpreted not in the context of our happiness but in the context of our unhappiness; a dream I had at the time of our first serious quarrel. I have not selected these dreams to write about arbitrarily, but deliberately because I want to use them to develop the thought that good dreams contribute to social integration while bad dreams are symptoms of dysfunctional relationships, so if we want to think about any vital problem whatever we need to think about dreams; and in general I always think carefully about how to begin, how to continue, and at any point on any page what word to write next, always trying to perform responsible speech acts; so now I have a plan in mind when I tell you our quarrel made me an unhappy person and that in my sorrow I sought consolation in religion.

On the nights after I prayed with my Catholic charismatic community my dreams reversed their bitter downward spiral into fear and anger and began spiraling sweetly upward into confidence and forgiveness on the wings of the hallelujah shouting, the speaking in tongues, the swaying, the embracing, the storytelling, the praise and the gratitude of the brethren migrating with the heavenly hosts *en masse* on flying horses over the plains of Israel to the warm comforting waters of the River Jordan where we bathed together singing like best friends singing Happy Birthday in a Jacuzzi. Unlike dreams that compensate in fantasy for frustrations in reality, these two happy dreams of mine continued pleasures I was already enjoying awake, as people who do sex on drugs trip out on the drug when they are already tripped out on the sex. My own experience with drugs has been mostly vicarious, but I have had a great deal of experience with religion, which I have found to be a healthier habit, a safer high, a wilder trip, and cheaper. My druggie friend argues with me that his is the more honest and decent way of life because doing religion involves believing lies and repressing people who just want to have fun. We disagree, but we understand each other's deep emotional needs. My happy dreams echoed the happy fulfilments of deep emotional needs in my waking life; while more commonly among human beings on this planet at this point in history there is a vast emptiness in the heart due to lack of satisfying relationships; an emptiness at the level of what Sigmund Freud calls the primary processes, at the level of dreams, at the level of the non-conscious mind (which is the greater part of the mind); which is why all around the world there are millions who are like I was hanging in the basement of the parish of *Notre Dame du Chemin* singing hymns with the brethren; and millions who are like Pablo the kid down the block who sniffs glue and steals avocados to pay for it. The subterranean dynamites of the non-conscious emptiness of millions of people, often detonated by humiliation, explode repeatedly as insanity and violence far worse than the mild nuttiness and naughtiness of me and Pablo.

Although one might agree that dreams express the strivings of the non-conscious mind, and one might agree that the greater part of the mind is non-conscious as are the hotter fires of motivation; and one might go on to agree that the quality of one's dreams is an index of the quality of one's life and life-energy; and consequently one might believe my schoolteacher friend Leno Venegas when he says that the students who study and learn are the students who have a dream, while those who are apathetic and destructive lack a dream to inspire them; and one might believe me when I report that love improved my dreams; one might nevertheless balk at believing me when I report that my participation in atavistic ceremonies improved them. It is perhaps easier to believe that the dreams of Little Hans were happier after Freud cured his fear of horses; or that the Wolf Man, who as a child had nightmares where he faced the cold white stares of six wolves in a tree, had good dreams after Freud deciphered the true causes of his anxieties.

Let us consider, first from a psychological and then from a sociological perspective, a dream of Freud's patient Dora, the eighteen year old daughter of a rich man Freud had earlier treated for syphilis, whom her father brought in for treatment because of her hysteria and a recent suicide threat, as well as migraines, coughs, abdominal pain, shortness of breath, gastric pains, loss of voice, fatigue, depression, and hypochondria. Dora dreamed three nights in a row, and then on a later occasion once, that her house was on fire, that her father stood at the foot of her bed to wake her up and save her, that her mother wanted to rescue a jewelry box from the flames, and that her father said forget it I don't want my children incinerated because of a jewelry box. Dora's conscious mind does not know until Freud's psycho-analysis that her dream is trying to protect her from her repressed desire to give in and let herself be seduced by a Mr. K who in real life really is trying to seduce her, by rekindling her love for her father, calling him to help her to flee from a place where she is in danger, as in her infancy her love for her father and his disciplinary presence had protected her from wetting her bed. The dream came to Dora when she and her father were guests at the lakeside country house of Mr. and Mrs. K, on the night after the day when Mr. K took advantage of a stroll with her along the lakeshore to declare his love for her and to steal a kiss, and he had apparently also stolen the key to her bedroom making it impossible for her to lock the door and sleep safely at night. Dora's dream expresses her decision to flee back to Vienna with dad; it is repeated nightly while her decision is not yet implemented; the dream fulfills in sleeping-life the wish that he be the good dad that in waking-life he is not, for in reality her father and all her family have treated Dora's accusations about Mr. K's earlier advances as sordid fantasies produced by her allegedly oversexed mind; in reality her father secretly wants K to succeed in seducing Dora in order to assure K´s acquiescence in his own ongoing affair with Mrs. K. Later Dora told Freud of a second dream: she is in an unknown city walking unknown streets when suddenly she is at her own house where she finds a letter from her mother waiting for her saying that since she had left home without telling anyone her parents had not told her that her father was sick and now he is dead and she can come if she wants to; so Dora (in her dream) starts for the railway station and asks a hundred people where it is and they answer "in five minutes" except for one who answers "in two and a half hours," and then suddenly she is at the building where her parents live and she asks the concierge for directions to their apartment, and is answered by the housekeeper, "They are all at the cemetery." This second dream falls in the category of dreams about the death of a loved one, which is one of three kinds of dream Freud mentions in the *Traumdeutung* as "common dreams," the other two kinds of common dreams being those about being naked in public, and those about being unprepared for a final examination; and surely Freud is right to think that from the perspective of building up a science of psychology or of psycho-analysis the prevalence of certain common dreams, as well as other less frequent but still familiar kinds like Dora's dream of being in a burning building, can be seen as evidence of the deep non-conscious action of what Freud calls affective primary processes reminiscent of deep-lying tectonic plates.

I think Plato would approve of Uncle Sigmund's efforts to bring more of the contents of the unconscious and preconscious minds into consciousness, for if we can (as I think we can) equate Freud's ego (*Ich*) with Plato's rational soul (*logistiche psuche*) then the two thinkers have a similar idea: the conscious self has the job of directing and harmonizing the emotions; so they should agree that it can do its job better if it knows more about those emotions. My friend Plato would say that dreams are invaluable because they display in codes that can be decoded feelings driving action the *logistiche psuche* hides from itself when it is awake. Plato would, furthermore, not be surprised by Adriana Aron´s recent study showing that over half of the refugees fleeing the Salvadoran Civil War have recurring dreams that someone is chasing them trying to kill them, because he realized long ago that what happens in the *psyche* reflects what happens in the *polis*. Nor would Plato be surprised when Hannah Decker shows Dora's dreams to be those of a typical "hysterical" woman oppressed by the patriarchy of old Europe who makes herself sick to get some semblance of control over her life, because he would expect psychic structure to reflect social structure. But it would perhaps be more Aristotle, my dear friend Aristotle, who observed that Athens was a democracy because its defense depended on its navy and therefore every citizen had to wield an oar to row its boats-of-war; who observed that city-states dependent on cavalry for defense were not democratic because the only citizens with military significance were the ones rich enough to keep horses; who would have observed if he had lived in Vienna in 1900 that Freud's patients were from a social class where men—as Karl Marx unkindly remarked with a certain perhaps pardonable degree of exaggeration—accumulated wealth exploiting the working classes and then devoted their leisure hours to seducing each other's wives; who with his talent for sociology would have observed how in Dora's Vienna the big picture (the society) matches the little picture (the individual); the waking life matches the dreaming life; as Dora spends her days caring for the children of the same Mr. K who is plying her with expensive gifts and soliciting her body because Mrs. K is out having fun with Dora's father; as Dora spends her nights dreaming she is lost and when she asks the way to the railway station people reply "in five minutes" or "in two and one half hours." In the terminology of Emile Durkheim who found in his empirical studies that rich people, poor people, and people without faith are more likely to commit suicide than middle class people and people with faith; who drew a map of Europe charting more suicide in the more modernized areas and less suicide in the more traditional areas; the confusion about roles and their duties that Aristotle perhaps would have observed in Dora's family and its milieu would be termed "normlessness" (*anomie*) and "social disintegration." If we graduate to a sociology of dreams and if we acknowledge happy dreams fulfilling wishes that are also being fulfilled in waking life, in love and/or in the loving support of an extended family, then it becomes easier to credit my report that I dreamed of migrating with the heavenly hosts gliding *en masse* on flying horses over the plains of Israel to the warm comforting waters of the River Jordan where we bathed together singing like best friends singing Happy Birthday in a Jacuzzi. I want to call it a dream of social integration; and I want to make it an example of a successful dream, a dream that marks the success of well-functioning institutions, as integration marks the success not just of Freud's therapy, but also of rituals performed by shamans in indigenous cultures, of therapies based on behaviorist psychology, and indeed of any therapy, as has been brilliantly argued by the Johns Hopkins psychiatrist Jerome Frank in *Persuasion and Healing*; and I would die happy if I could persuade more people to read Frank's book and if I could communicate all I want to say with the phrase "dream of integration." I want to say "integration brings happiness, loneliness despair" with Durkheim and "a dream stands on two legs, one present, one from early childhood" with Freud. I do not know whether Freud's "two legs" theory is true for all dreams. But I believe that its two parts, one a generating event in the present life of the dreamer, the other a rekindling of primary erotic experience, help me to interpret my two bonding dreams and to explain "dream of integration." I believe the same early childhood erotic joy is rekindled in my love dream and in my prayer dream, and that its source is being rocked back and forth in my mother's arms held close to her warm body sucking warm milk from a nipple of one or the other of her breasts. I also believe the joys of flying and of warm bathing in these dreams (and my other similar ones) are enhanced by feelings of relief from fears of abandonment and attack that I acquired later in childhood and youth; they are homecoming dreams. I am one of the lucky ones who has a home to return to in dreams, as Odysseus returned to Ithaca to Penelope and a "welcome" ("welcome" translates Homer's word "*agape*" -- a word which later named "love" and "God" in the New Testament); I am one of the lucky ones who as a baby learned Erik Erikson's "basic trust." I believe my basic trust in early childhood was one leg of two happy dreams later, one sparked by love and the other by prayer; and that these good dreams illustrate a connection psychologists already believe to exist between infantile bliss and ability to form healthy relationships in adult life, which in turn suggests a more general connection between the constructive channeling of erotic and other primary emotions (Freud would say their "sublimation") and social integration.

Many year's after I stopped sucking milk from my mother's breasts, many years after I stopped sucking my thumb, I had an opportunity to discuss John Bowlby´s book *Child Care and the Growth of Love*, a book which amasses evidence to show what I believe common sense already knows: that babies who are not touched, kissed, and cooed over are likely to grow up to be unhappy, dysfunctional and anti-social, with Professor Lawrence Kohlberg of Harvard. Kohlberg suggested that people not loved as babies would be unable to move from the stage of obedience to authority to the stage of wanting to be nice girls and nice boys appreciating others and being appreciated by others, in other words unable to move from Stage Two to Stage Three on his scale of moral development, because they lacked an urge to come home to infantile bliss. I conclude that since without passing through Stage Three nobody goes on to a higher stage, it follows that a nation of unloved babies must necessarily become a nation of irresponsible citizens. This conclusion is buttressed by all I know from other sources, including my own trite observation in the classroom that the children who want to please their parents are the children who want to please their teachers; bearing in mind the need for confirmation from other sources because Kohlberg is open to the objection that he is biased in favor of the male and the rational; and because Freud (at least prior to 1920) is open to the objection that for him any emotion at all is erotic —for example although Freud did not question the clinical evidence presented by his colleague Alfred Adler to support the claim that aggressiveness and will-to-power are the mainsprings of human behavior, he reinterpreted Adler's cases, reading them as showing how frustrated love turns into aggressiveness and will-to power, reclassifying them as re-channelings of erotic impulses in destructive directions in ways reminiscent of Saint Thomas reclassifying evil as the absence of good, evil being for Saint Thomas failure to achieve the love that God intends.

Recently I checked for several days the "Trending Now" charts on Yahoo showing a day's ten topics most searched on the Internet. The most common searches were for sexy female entertainers, followed by sexy male entertainers. This modest foray into quantitative empirical research (using a methodology other inquirers can replicate) tended to confirm three of my opinions: first, although Uncle Sigmund no doubt overestimated Eros and defined it too broadly, he was not entirely wrong in making a scientific case for saying sex makes the world go 'round; second, Uncle Sigmund was certainly right on the broader point that we are often motivated by strong primary emotions we understand dimly if at all; and, third, that as long as five hundred or more fill a Pentecostal church, fifty thousand or more attend a rock concert, and one hundred thousand or more flock to a football game, while only six show up to discuss how to save humanity and the planet from certain destruction, humanity and the planet probably will not be saved from certain destruction.

Let me now write the rest of this chapter as an effort to make clear more of my opinions on dreams and how they figure in a Game Plan to get *homo sapiens* off the endangered species list, free of attempts to prove that my opinions are true, or original, or what Sigmund Freud meant. My goal is to contribute to reorganizing a modern world-system where hysterical women like Dora and paranoid men like my father are generated by dysfunctional institutions ill-suited to the human body as it has evolved over the millennia and ill-adapted to the surrounding biosphere; bearing in mind that myths (today most notably Smith's myth of natural liberty) organize human action and that dreams drive it. When we say "the dream drives the action" we mean not just a night-time experience during sleep but also the day-time motivation energized by the primary affective processes that the night-dreams express and reveal—motivation standardly organized by the *logistiche psuche*'s conventional norms at the point where the rubber (the myths) hits the road (behavior). Norms: to understand capitalism we need to understand its constitutive rules, its norms. Dreams: to transform capitalism we need to engage the underlying tectonic energies of our driving emotions (*Triebe*).

My goal is different from the goal of people who study when discontent articulated in radical ideologies will rise to the level where majorities see change as a possibility and cannot stand the system any longer and are able to pull off a successful revolution, because I believe the real transformation will be the shift in constitutive rules that frees humanity from the iron cage of the logic of accumulation, a shift which—as we should know by now—revolution does not guarantee. Escape from the iron cage is not separate from the constructive tasks of organizing green alternative ways of cooperation and sharing like (four examples) community supported agriculture, local currencies, gift economies, worker ownership. My goal is the same as Arnold Toynbee's when he concludes from his study of history that successful civilizations are ruled by charm, while unsuccessful ones are ruled by force because I agree with Freud that civilization is energized by the sublimation of *Eros*. Charm begets partnership, force domination (Eisler). Although I agree too with Herbert Marcuse that there is often a "surplus repression" of pleasure making people more gloomy and more fearful than is necessary to motivate them to work hard enough to make the economy churn out the necessary goods and services, I want to put a positive spin on his critique. I want an esthetic education leading to ethics (Schiller), I want to equate truth and love (Gandhi), I want beauty to prepare the soul for the entrance of right reason (Plato). To that end I will now suggest a thumbnail sketch of the social history of charm, beginning with the first family ties. The glue binding the first human groups was the women's sex appeal. The males hung with the women and children because—unlike other mammals—the human females were always in heat. (Tanner, Leakey)

Of course this account of the erotic origins of human kinship presupposes that there were already strong positive emotions bonding the women and their children; and we know too that later if not earlier diverse cultural resources fostered social cohesion. If the clans and tribes of our remote ancestors had not known how to achieve enough group cohesion to cooperate in food sharing and child care, they would not have survived. The ancient cultures that survived learned to turn on the charm in many ways to promote functional behavior and discourage dysfunctional behavior, including music, magic, feasting, dancing, singing, gift-giving, initiations, rituals, pageantry, totems, spirits, ceremonies, reverence for deities, going into trances, costumes, ancestor worship, sacred animals, sweat lodges, chanting, narcotics, holidays, games, contests, jokes, mysteries, secrets, stories—in short everything that is fun and exciting, deployed to achieve socialization into norms and roles—as we can see in the charms that have lasted into modern times.

To make the mirror image of the same point: among pre-hominids and hominids and then during the many millennia of human evolution the emotions themselves (the physiological arousal states) have evolved to lend themselves to being used by cultures in ways favoring the survival of individuals and of groups.

Which brings me back to our charismatic prayer group in Quebec where we made it a point to follow exactly the practices described in ancient texts; we even danced to rhythms and steps authorized by the Old Testament. I make bold to suggest the hypothesis that the explanation of the resulting improvement in my night life was that people long ago learned to bond with integrating charms nobody then or now fully understood or understands, and I suggest that I was the beneficiary of some of those tried and true charms when my dreams switched from anxiety to celebration; but I do not think for one moment that advocates of abstinence, virginity, and celibacy weaken erotic drives; on the contrary with Freud I believe that Eros tops the charts whether the ranking is measured according to how much or according to how little; and to demonstrate my point I call as a witness my friend Patrick, a Catholic taxi driver, who like Dante Alighieri believes in constant conversion, weaving one's eternity thread by thread day by day, hour by hour; and because what he means by constant conversion is constant purification from the temptations of sex, Patrick enjoys an exciting life cruising around town in his taxi fending off the temptations of the devil minute by minute.

The story of modernizing economic calculation leading to the disintegration of traditional kinship norms and the commodification of almost everything (and here ends my thumbnail history of charm, at the point where charm becomes a commodity) has been told often and well by historians like Karl Polanyi and sociologists like Max Weber. Nestor Garcia-Canclini adds that indigenous cultures still endure *malgré tout*; they share space with modernity in contemporary "hybrid" societies; or as Robert Bellah puts it now modern and ancient "languages" coexist. In the same institutions and even in the same person a modern head may share power with an ancient heart, and indeed more than one head with more than one heart. Nevertheless, I single out for emphasis that to solve humanity's problems the language of economics must be tamed.

To tame economics (the beast, the box, the basic structure) I now beg leave to suggest eleven thoughts. Although no reader is under any obligation to believe my thoughts, I can scarce imagine a reader doubting that they are important if they are true; nor can I conceive—blinded as I may be by my own eyes—that they be false, since they just restate facts established by others—as anyone can verify in any library—in a format designed for drawing practical conclusions. Following throughout Charles Taylor in construing our basic structure (the box) to be the constitutive rules of a bargaining society; and finding beginnings of our bargaining society two thousand five hundred years ago; I will suggest four thoughts about its history, two thoughts about its consequences, and five thoughts about a Game Plan for turning the tide with respect to humanity's and the biosphere's present predicaments. History: beginnings of our basic rules are found in Rome. Over time similar laws and ethics from elsewhere in Europe joined the main stream. The first laws served to prevent fights among *patres familias*. Feeding babies, cooking, and caring for the sick were not relevant. Those were matters for women and slaves; the purpose of the law was military: to avoid internecine fighting that would sap Roman power. So at the first start of law in Rome—I am not talking about Confucian, Hindu, or African law, which were more like community therapy—all the dream energy of bonding and all the charm energy of integration were marginal to jurisprudence. Second: emancipation (getting out from under the *mancipum*, the hand, of the *paterfamilias*) would henceforth mean becoming too a sovereign autonomous individual like him. Third: some seven hundred years later the Romans invented their law of nations, *jus gentium*, constituting a universal legal subject with a few universal rights; disregarding local kinship obligations, languages, religions, customs, ceremonies, roles.... Under the *jus gentium* commerce could be transacted anywhere in the Roman Empire under the same rules. Fourth: in the reception of Roman Law in early modern Europe and in British common law "contract" (Pufendorf 1688) came to be "a meeting of minds"—a definition which had always been implicit in the very idea of the sovereign autonomous individual; and then in Adam Smith contract became the principle of the division of labor, the principle of the market (Smith 1776). Europe changed from a "status society" to a "contract society." In a contract society—unlike a pre-modern *societé segmentée* that is in principle an extended family—where there is no contract there is no reciprocity of duties, no duty to share food for example; except of course when in hybrid societies non-mercantile logics (like those of family ties, statutory entitlements, divine authority...) complement the constitutive rules of a bargaining society. The consequences: A first consequence is exclusion; property is by definition exclusion. *Dominium*, meaning place conquered, was the ancient Roman word. *Proprietas* came into use as its synonym well after the Republic under the Empire.

Those who own no real estate and can pay no rent are therefore excluded from space. They have no right to be anywhere. A second consequence of the constitutive rules is the ungovernability of a world where rational action to solve problems is constrained because nothing can be done that violates for example the systemic imperative to prevent inflation; since for money to be a medium of exchange it must first be a store of value; and in a market society livelihood depends on exchange; so therefore inflation must be curbed even at the cost of unemployment and cutbacks of social programs. I had a recurring dream about ungovernability—about a lack of capable authority—starting when I was four years old: in my dream I was a passenger in a car speeding along a road, and then suddenly I realized to my horror the car had no driver; the driver had disappeared, there was only an emptiness in the driver's seat as the car careened to destruction. This dream started after my father was laid off from his job. I was too young to understand Keynes, but I could already guess that the man who does not get the money does not get the girl, and that since my father was losing the money he was likely to lose the girl, and I a father.

After Deleuze and Guattari I need not apologize for calling the empty driver's seat not a wish but a fear, but perhaps I still must justify my psychological cure for an economic disease.

The psychological cure requires basic trust in early infancy as a prerequisite for dreams of integration and adult moral development. The economic disease is a chronic shortfall of effective demand that makes unemployment and instability chronic too. The former cures the latter by laying foundations for integrated communities where people have dignity and livelihood even when it would not be profitable to hire them.

I shall attempt to summarize this chapter by telling about my recurring dreams of swimming in swift blue wide torrential rivers dragging me downstream in the direction of rocky Niagaran cataracts, while I swim sideways as hard as I can to incline my trajectory left toward the riverbank, racing to reach the shore before the falls. I interpret the raging river as the systemic imperatives that now rule the modern world-system. But I make it to shore, and then I am in the secret hot tubs of Skull and Bones at Yale, in the Turkish baths at Valparaiso, in room after pleasant room of waters rechanneled to cater to my every desire. The warm waters that bathe me, change my diapers, dust me with talcum powder and cover me with kisses, represent my safe arrival at the other economy that is possible; they welcome me and all the towel-clad steam-surrounded bathers at the Turkish baths, and all the Yalies and all the townies, home to the fraternity that every cell in every body yearns for; they meet all our needs in sustainable harmony with the environment.

Six

The disjointed vagary of his eyes shoots fear into my heart and mind. In his eyes I am a rich foreigner; while he is trapped in unemployment; I fear becoming a victim of rage born of humiliation. Thinking it wise to make friends, I ask him his name and what he thinks of the Green Life Center, where he has come for a free meal and where I am studying welfare in Argentina. His name is Vicente and what he thinks confirms Machiavelli. As that Italian sage would have predicted, Vicente prefers not to talk about the benefits he receives but instead about the benefits he confers: he does maintenance work on the building, cleans up rubbish around the neighborhood, and passes out flyers to announce Center events. Meanwhile lady volunteers from the neighborhood are slicing squash and dicing onions and carrots to throw into a single soup boiling in a single cauldron heated by a single stove-iron circle of flame fueled from a container of liquid natural gas purchased –as are the vegetables and a little meat added for flavor—with a four hundred dollar a month subsidy the Center receives from the City, which is all the cash from outside the Center gets because the Province contributes only beans, rice, lentils and other dry goods, while the Nation and the Private Sector do not contribute anything; and consequently there is no free meal served every day, but only a free meal served two days a week. I do not want to ask Vicente whether he knows that the philosophy of the authorities of the City and the Province calls for further reducing the number of free meals per week from two to zero because although public support for the nutrition of the poor was justified in the emergency of 2001, now that the unemployment rate is hovering around 8% according to official figures, the poor ought to go to work and earn their bread; nor do I want to ask him whether he knows that among some four hundred meal centers in the Province nearly fifty have already been shut down. Evidently the authorities are not reading—or if reading not believing—John Maynard Keynes' theory of liquidity preference; evidently they believe that structural unemployment is an unusual phenomenon that normally does not occur. They do not agree with Keynes that there is a chronic and normal weakness of effective demand, and hence a chronic normal tendency for there to be more people who need to sell something to make a living than buyers. Their policy is to shut down meals at Centers, and then to spend the same budget now spent on meals to give each indigent enrolled to eat a small one-time grant to start a micro-business. In their paradigm sharing food is unnecessary in normal times. The authorities are evidently disregarding Martin Luther King Jr.'s aptly titled book *Where Do We Go from Here: Chaos or Community?* where King proposes employing people in community service for pay, in caring for the old, the young and the sick for examples; and calls on us who own the wealth to pony up the cash to pay for it in order to build community and stave off chaos, assuming of course that most people will not do community service for pay because their livelihoods will come from other sources like for example wages, micro-credit schemes, public employment, professional fees, living in monasteries, or juggling on stilts at stoplights for tips. King's idea is to use employment in human services as part of a mix that adds up to the social and economic integration of every sister and brother, but apart from calling for justice he does not propose a specific way to pay for it.

I have an old-fashioned opinion about where to get the money to pay people like Vicente to do useful work (and to go to school to improve their skills and lives) without necessarily having to find paying customers to buy their services or their products, which goes back to Adam Smith' s (1776) (and David Ricardo's) idea that a tax can be levied on three possible sources: on wages, on profits, or on rents (now we can say on the value added by labor, the value added by capital, or the value added by land); to which we might now add value added by the creativity of entrepreneurs (per

111

Joseph Schumpeter) and the value added by knowledge (per Peter Drucker). Although I am drawing on Smith's ideas about taxes, I am not just talking about how to collect taxes, but about any way to channel money to Vicente; however the money may be channeled I think I need to consider whether it comes from wages, profits, rents or another source.

Here I want to generalize a point Adam Smith implies and David Ricardo (1817) makes explicit: rents (rather than profits or wages) are the best source for taxes—a point I generalize to say that money from rents (in the ricardian sense where the word "rent" means earnings from owning natural resources like especially fertile farmland, forests, and minerals) is an excellent source of funding for community service. Rents can be channeled to pay for Vicente's community work via tax-and-spend, but also via donations from people or institutions with incomes from rents; and of course Vicente's employed neighbors may help too via fundraisers asking them to give part of their wage income, and of course businesses may chip in with some of their profits. A beauty of this marriage of Martin Luther King Jr.'s proposal for making employment unlimited by meeting all the staffing needs of the human services, to the classical ricardian and Fabian principle of socializing rents, is that it builds conditions for raising the wages of the working poor, because full or nearly full employment stokes the leverage of underpaid workers bargaining for higher pay and better working conditions. Vicente should appreciate the importance of broad society-wide increases in poor people's incomes because when free meals at Green Life end and he gets his chance to be a micro-entrepreneur, then (studies show) even assuming his micro-business succeeds (less than half last two years) his monthly net is not likely to be more than eight hundred Argentine pesos (about two hundred US dollars).

I hope nobody will complain that they cannot understand these ideas because I am tired of being told I am brilliant but nobody can understand me and they are perfectly simple; there is nothing off-the-wall or impractical about them: you work but your pay does not come from sales; the pay can come from anywhere but an especially recommended source is revenue generated by natural resource rents; which is perfectly straightforward because people are doing community service work all over the world every day, and every day governments and private owners are raking in billions from petroleum. Petroleum is just the beginning since other natural resources also generate incomes for their owners (whether the owners are private or public) far beyond any expenditures needed to motivate and accomplish production; and the principle extends even beyond natural resources because, for example, if a football player is willing to play just as hard and just as well for two hundred thousand dollars a year, but because he has a good agent and because the sale of television rights creates a bonanza in the football business he rakes in two million a year, then most of his income is "economic rent" according to present-day post-ricardian definitions of the term. Finding the money to pay for King's full-employment-through-human-services proposal is an economic problem easily solved compared to the educational problem of facilitating a culture shift toward a work ethic and toward a self-improvement ideal, among people who for generations have lived between crimes and crumbs, between violence and humiliation; getting by on burglary and begging, protection rackets and government largesse, drug-dealing and ass-kissing—a culture shift which I think we can assist by listening to Plato's suggestion that all education should begin with music, and drawing the conclusion that it is a step in the right direction for the kids in the slums to learn to dance and to play an instrument.

There are already examples of recycling ricardian rents to fund community service and some of the examples are in Argentina, like the famous "work plan" program of former president Nestor Kirchner that hired unemployed parents to do community service in exchange for a small monthly subsidy from the government—subsidies paid for mainly by the famous "retentions" taxing exports mainly of soybeans and grains mainly to China and Europe. This taxing of exports can be considered a taxing of ricardian rents because their source is revenues from exploiting a major natural resource, namely Argentina's fertile *pampas*. Of course I know it can be objected that full employment, whether achieved by policies including Kingian human services or in some other way, would be a curse because it would cause inflation, and I know it can be objected that high wages would be disastrous because they would force marginal employers out of business. I have considered these objections: indeed I use the so-called trade-off between inflation and unemployment, and the necessity in many cases to keep wages low, as reasons to call for a paradigm shift; and as motivations for thinking up a Game Plan for moving from the old paradigm to the new one. Within the dominant paradigm there is a discouraging tendency for the solution to one problem to be the cause of another (for example for the solution to the problem of inflation to be the cause of unemployment); and in the academy a discouraging blinding tendency to tell students they have not properly identified their research problems until they have isolated one problem from another, unemployment from inflation, low wages from crime, drugs from mental illness, the burning of the rain forests from the global economy, while in fact the governing paradigm has us in an iron cage; and in politics a discouraging blinding tendency to identify being realistic with addressing one issue at a time, while in fact the governing paradigm has us in an iron cage; and in the worlds of business and public administration a discouraging tendency to identify rational management with establishing specific measures of the performance of specific tasks, while in fact the governing paradigm has us in an iron cage; and in the face of all this blind discouragement I recall an old Baptist hymn from my youth "this little light of mine

I'm gonna let it shine" and—inviting your comments—I outline a Game Plan for changing the paradigm, not so much by refuting neoliberal theories, as by changing the facts in the real world that give neoliberal theories their credibility. As a bridge to a new paradigm I propose to defang the tendency of full employment to cause inflation by taxing the rich and pluralizing production, agreeing with the monetarists that inflation is too much money chasing too few goods, and therefore curbing inflation by decreasing money and increasing goods: taxing the rich (or somehow persuading people to hide money under mattresses or otherwise take it out of circulation) to lower the amount of money, and pluralizing production to raise the amount of goods. Although I generally downplay conspiracy theories, I nevertheless admire the unraveling of conspiracies to raise prices achieved by Nestor Kirchner's Minister of Economy Felisa Miceli who brought consumer, distributor, and producer representatives to the same table to examine the facts; who forced businesses to disclose supplies of goods and their costs; thus discouraging the secret hoarding of goods to drive up prices; and thus unmasking the bogus legitimating of price increases by exaggerating costs. But admirable as conspiracy-unmasking is, inflation will not cease to be a threat without a paradigm shift—a topic I am leading into with my proposal to take money out of circulation by taxing the rich, a proposal that inevitably triggers the argument made by Republicans in Congress, by Friedrich von Hayek, Milton Friedman, and many others that taxing the rich will slow down investment, job creation, and production; worsening inflation by decreasing instead of increasing supplies of goods; and I am saying the problem is not that Republicans make this argument: the problem is that their argument is true. I am saying that the first part of my modest proposal (taxing the rich) is only feasible together with the second part (pluralizing production). I say refute their argument by changing the facts that make it true.

Because there are many ways to make it false that a fairer distribution of wealth and opportunity brakes production, a fairer economy is plural; the logic of capital accumulation may be one of its logics, but must not be its only logic. But many Argentines will object that President Nestor Kirchner's work plans already did what I am recommending—pluralizing production by paying people with public money to do community service—and the result was corruption and chaos. Many Argentines will object that all or most of the Vicentes with work plans either did not do any community service at all, or they did only one service, which was voting for the candidates endorsed by Kirchner's political machine. Argentina's bad experience with the logic of the work plans partly explains the provincial government's backing of the logic of entrepreneurship; it argues for pushing micro-credit and micro-enterprise for educational reasons: to instil a work ethic and to foster a culture of responsibility. My conclusion is that the paradigm shift required to make the Republican argument false must both solve Keynes' problem, the problem of compensating for a chronic weakness of effective demand by creating work that does not depend on finding customers to buy the products (which the work plans solved); and also solve the cultural problem Joseph Schumpeter identified when he said institutions are like battleships: they must be properly built *and* properly manned (which the work plans did not solve).

I am only asking for a little realism and a little imagination; realism to face the fact that no combination of capitalism with bureaucracy can ever create community; imagination to dream *ubuntu*, integration, sisterhood and brotherhood; realism to face the fact that markets must be restrained and supplemented by non-market relationships, many of them non-mercantile (Amartya Sen). Your dear heart may tell you that nothing would be more peachy-keen than a planet without boundaries, chemically-pure free trade, where capital, goods, and people circulate freely around the globe governed only by the "natural" laws of commerce; but I want your brain to tell your heart that such a world, peachy-keen in Smith's eighteenth century new-birth-of-freedom heaven, is hell on twenty first century earth. I want your heart to tell your brain to think outside its rut: think human relationships, think not just one metaphysics but many; think not one mythology but many; think the earthist challenge to economism (Cobb). I request two more realisms: first realism to face the necessity

118

of the magic of myth, because without myth there is no community, and without community there is no such thing as a human being; because there is no physical survival without social order, no social order without norms, no norms without culture, no culture without myth, no myth without dreams; second realism to stop looking for someone to blame every time a nation cannot pay its debts, or a bank collapses, or when there is mass unemployment or runaway inflation or unending poverty, or whenever years of sullen resentment boil over into mass violence; as if it were caused by somebody who overspent or failed to regulate the banks or in some other way violated the norms of good government and sound economics; as if ours were a normal world where correct application of normal rules leads to peace and prosperity; as if every disaster were caused by somebody's error; when in fact in a world organized by the basic rules of our so-called "civilization" disaster is normal.

Imagine permaculture, fair trade, worker-owned businesses, extended families, barter networks, and conscious consumers. Imagine common security clubs (Chuck Collins), gift economy, gleaning, Ithaca hours (Edgar Cahn). Imagine a profusion of practical logics; corresponding to what Max Weber calls *Zweckrationalität* (goal-oriented rationality), think of a logic for each goal; one for the iron goal of capital accumulation; one for *quebecois* dairy farmers organized in cooperatives whose goal is to keep their farms and families going—not to turn cash into more cash; a third for the members of the third order of Saint Francis whose goal is perfection in charity; and so on; while corresponding to Weber's *Wertrationalität* (customary rationality) think of a logic for every custom, judging every action by the requirements of a norm, or by the duties of a role: the logic of being a good doctor, a good father, a law-abiding citizen, a responsible consumer who follows the norm of reduce, reuse, recycle, and so on.

Here is the heart of the Game Plan: it is not a matter of deducing from theory what to do and then doing it; it is a matter of identifying the growth points in the existing culture; as Methodists say, it is discerning God's work in the world and then joining it. The sum of many growth points will empower us to say: "we want the good businesses that respect unions, obey the law, serve their customers well, use their profits with social conscience, and help build a green and frugal future (the only possible future), but bad businesses we do not need because now if they fold nobody will be hurt; other sectors of our plural economy will take up the slack, assuring livelihoods for their ex-workers and ex-managers." We empower ourselves to require that business be socially responsible by using multiple logics: like those guiding indigenous practices like *minga* (Latin America) *ubuntu* (Africa) and *sarvodaya* (Asia); and cooperatives, barter networks, solidarity economics, and of course micro-credit and community service for pay. By building a strong, honest, and efficient public sector, financed not only by taxes. By winning some degree of autonomy from the competitive pressures and neoliberal rules of the global economy.

Some of my business-owning friends may demur that it is harsh to close an undercapitalized marginal business, liquidate its assets and pay the proceeds of liquidation over to its creditors, just because it was five years behind on taxes, or just because it ran a fleet of uninsured trucks and one of them ran over a little boy and it could not pay the tort judgment, or just because it was paying its employees less than minimum wage; for, after all, it is legitimate self-defense to do what you have to do to survive. I reply that when a business is liquidated, the owner does survive; it is not a homicide—all the actors in the drama are still standing on stage afterwards—and the employees and managers who lost their jobs can be paid to go to school while they are unemployed. This kind of solidarity with the human survivors of liquidated legal fictions has been standard practice in Sweden since the 1960s, but paid transitions there and elsewhere have come under pressure lately along with all of welfare state social democracy. They need a paradigm shift to work reliably. New thinking is needed to move beyond a principle Keynes considered obvious, namely that employment is necessarily limited by market demand, and until that principle is rejected we cannot reliably assure people who lose their jobs that soon another slot in the economy will open up for them. This is not hard to understand or complicated and I am tired of people who admire me for being smart and well read while they act as if I had never written a single word. It is not hard to understand that in the dominant paradigm—inside the logic of the market—no policy works. Sen makes my point so clearly that it is hard to misunderstand. He says that the logic of the market must be restrained and supplemented by other logics, many of them non-commercial.

It follows as the day the night—somebody please correct my mistake if I am missing something—that we ought to be practicing those other logics, matching other thoughts with other deeds, cooperating and sharing above and beyond working for pay and investing for profit. Mahatma Gandhi made this point in simple words and numbers when he wrote that socialism begins with one person, when one person starts to behave as a socialist. One can become 10, and then 100, and then 1,000,000; but if the initial number is zero, then however much it is multiplied, the sum will be zero; therefore besides being realistic enough to see that neither justice nor peace nor sustainability can be expected within the dominant paradigm, and besides being imaginative enough to understand the dairy farmers of Quebec and the eye doctors of India and the millions of others who march to the beats of different drummers, we ought to be responsible enough to practice an ethic higher than the dominant ethic. Right wing economics is not entirely wrong as academic theory and it is best refuted by changing the facts it assumes, by following logics different from its logic, thus building full employment in a green and frugal economy with a stable currency by creating left wing and centrist alternatives in practice.

For several decades now the Volker Fund, the Olin Foundation, and many other funders have been paying scholars to seal off every possible intellectual and scientific exit from the iron cage. But the key problem is not that setbacks at the levels of academic politics and academic financing have thinned the ranks of maverick social scientists; the key problem is that in real world experience to the extent that me getting a job depends on someone making a profit by hiring me, employment depends on profit-accumulation (Aglietta); and therefore the key solution is to construct regimes of non-accumulation—which is in theory feasible because as Piero Sraffa shows in *The Production of Commodities by Means of Commodities* while to produce goods it is necessary to take the technically required actions, the actors who take them need not have the motives assumed by mainstream economic theory, but could have any motives at all; and which is in practice feasible because whatever industry you look at—schools, hospitals, oil, sports teams, air lines, television, dairies, retail, banks, child care, insurance ...—it is evident that nonprofit, public, indigenous, voluntary, worker-owned and/or cooperative firms can produce and distribute goods and services.

I am trying to capture the common phrase "think outside the box" (which is already a popular idea) to make it mean being able to imagine employment for Vicente—or, to be more precise to imagine Vicente's

123

social integration—outside modernity's main dominant institutions, which are (following Max Weber) capitalism and bureaucracy. I am advocating a gestalt shift —Freire would call it a consciousness-raising—which re-envisions "the box" as tribal myths, which sees contract law and property law as a tribe's truncated ethics, as the causal matrix of its nightmares. I want Argentina and every other country to abandon once and for all the illusion that modernizing will bring full employment at good wages. I want to hinge "thinking outside the box" to building a plural economy where business for profit is just one sector.

I could say also "economy of solidarity" or "green economy of solidarity." What I like to call the "love sector" is an indispensable part of the "solidarity"—it need not be all of it, but it has to be part of it—because supplementing private-business-for-profit with cooperatives, worker-ownership, public ownership and so on runs up against the barrier that whoever owns the business and whether or not it makes a profit, it still has to balance the books. The "love sector" includes volunteering, gift-giving, sharing within families, sharing within clans, social entrepreneurship, parenting, grandparenting, and all the mutual aid prescribed by culture that Alvin Gouldner and Karl Polanyi call "reciprocity"—and far from being a minor aberration reciprocity is (more than any other economic form) the norm in the history and prehistory of the human species. Let me give the example of us and our neighbor Pedro. From the red grapes that grow on his grapevines, Pedro makes a sweet-wine brew called *chicha*. Most of the year we give Pedro lemons from our lemon trees. Once a year Pedro comes to our door with gifts of bottles of *chicha*. We do not calculate the market value of the lemons we give him and he does not calculate the market value of the *chicha* he gives us. Among indigenous peoples such arrangements have in some cases continued for thousands of years, with one family forever giving seafood and another forever land-food; and they can be and often have been sanctified by myths and ceremonies. Now compare Pedro with William, the owner of an insolvent wholesale fruit and vegetable business, who was my client when I was practicing law in Santa Barbara, California. William (facts and name altered to preserve attorney-client privilege, but the point the same) confided to me that if I could just help him get reorganized under Chapter Eleven, then he would promise never never never to let his business be "upside-down" again. By being "upside-down" William meant owing accounts payable exceeding his accounts receivable. While Pedro's game is sustainable, William's is not, because one person's payable is another person's receivable; therefore the sum of receivables equals the sum of payables; therefore it is impossible for everyone to get what William wants, namely receivables exceeding payables; someone must become insolvent, someone must be "upside-down" and indeed in the USA lately about a million people a year have been filing bankruptcy. I cannot answer all questions here but I can recommend my conclusion. A love sector is indispensable because the basic buying-and-selling game excludes many; and while government, cooperatives, micro-enterprises, worker ownership etc. do alleviate exclusion, we still need the buffer of a love sector to reach one hundred per cent inclusion –a conclusion that will be resisted by cool optimists who think it sufficient for governments to include whomever the market excludes. The cool optimists should read *The Dilemmas of Social Democracies* by me and Joanna Swanger. Note also: I do not just mean getting money to poor people. The excluded need enchanting dreams, dignity, sisterhood and brotherhood the same as you and I.

Now please try to imagine a high wage nation in a global economy. Here again we can learn from the Swedes. Imagine a feature of the old Swedish model of the 1960s: analyze an economy as two economies; where the larger is the national non-tradable economy (jobs like kindergarten teacher) where (assuming barriers to immigration) international competition does not affect wage policy; and where the second and smaller is the tradable economy where the nation is a price-taker—where the prices of its export products are set by international markets. Facts: sometimes (consider Chilean copper and Argentine soy beans) the price of a product exported is so high that the industry can afford high wages, and the industry cannot move because it depends on fixed natural resources; while in other cases (like the case of Volvo building plants in low-wage Brazil) the industry can move and policy fails to anchor it to a territory. Sometimes the product's price set by the global market is low. Then if one wants to continue in that line of export business at all one may have to subsidize wages to be able to compete, and one may subsidize, for a culture of solidarity allows subsidies.

My suggestion is that in a plural economy not desperate to obey the systemic imperatives of the logic of accumulation steps can be taken to promote reasonably high wages.

Indeed steps can be taken to promote whatever is desirable or necessary—but getting to such an economy requires a culture shift. It requires a shift not to one new myth but to diverse processes of "we" building, of reality testing, of collaborative problem-solving. Remember that Dewey's "experimental society" and Popper's "open society" —two great unfeasible visions I want to make feasible—were not fixed blueprints, but methods for getting feedback from experience.

I have tried in these pages to build "we" by encouraging practical cooperation; I affirm a larger "we" still under construction in the cheerful traditions of Bloch's Principle of Hope; and also of Paul's principles of faith, hope, and love, while in my own way agreeing with Nietzsche that Paul is Platonism for the masses. I know I am an odd bloke who has arrived at his odd opinions through odd personal experiences. But on my main points I am unable to imagine how I could possibly be wrong—which of course might be a weakness in my powers of imagination more than as an objectively existing impossibility. Still, although all truth be in principle tentative, and although it be pretentious to claim to have a message other people need to hear, I do not see how I could be mistaken when I say that Vicente and millions of women and men in situations like his will be marginal to society and at risk of becoming violent, or mentally ill, or drunkards, or dealers in drugs or pimps or worse, as long as employment depends on sales.

www.ingramcontent.com/pod-product-compliance
Lightning Source LLC
Chambersburg PA
CBHW022339280326
41934CB00006B/690

* 9 7 8 1 9 3 7 5 7 0 0 1 9 *